P, Malt

GUIDELINES FOR
SAFE PROCESS
OPERATIONS AND
MAINTENANCE

Publications Available from the
CENTER FOR CHEMICAL PROCESS SAFETY
of the
AMERICAN INSTITUTE OF CHEMICAL ENGINEERS

Guidelines for Safe Process Operations and Maintenance
Guidelines for Process Safety Fundamantals in General Plant Operations
Guidelines for Chemical Reactivity Evaluation and Application to Process Design
Tools for Making Acute Risk Decisions with Chemical Process Safety Applications
Guidelines for Preventing Human Error in Process Safety
Guidelines for Evaluating the Characteristics of Vapor Cloud Explosions, Flash
    Fires, and BLEVEs
Guidelines for Implementing Process Safety Management Systems
Guidelines for Safe Automation of Chemical Processes
Guidelines for Engineering Design for Process Safety
Guidelines for Auditing Process Safety Management Systems
Guidelines for Investigating Chemical Process Incidents
Guidelines for Hazard Evaluation Procedures, Second Edition with Worked
    Examples
Plant Guidelines for Technical Management of Chemical Process Safety, Rev. Ed.
Guidelines for Technical Management of Chemical Process Safety
Guidelines for Chemical Process Quantitative Risk Analysis
Guidelines for Process Equipment Reliability Data, with Data Tables
Guidelines for Vapor Release Mitigation
Guidelines for Safe Storage and Handling of High Toxic Hazard Materials
Guidelines for Use of Vapor Cloud Dispersion Models
Safety, Health, and Loss Prevention in Chemical Processes: Problems for
    Undergraduate Engineering Curricula
Safety, Health, and Loss Prevention in Chemical Processes: Problems for
    Undergraduate Engineering Curricula—Instructor's Guide
Workbook of Test Cases for Vapor Cloud Source Dispersion Models
Proceedings of the International Symposium and Workshop on Safe Chemical
    Process Automation, 1994
Proceedings of the International Process Safety Management Conference and
    Workshop, 1993
Proceedings of the International Conference on Hazard Identification and Risk
    Analysis, Human Factors, and Human Reliability in Process Safety, 1992
Proceedings of the International Conference/Workshop on Modeling and Mitigating
    the Consequences of Accidental Releases of Hazardous Materials, 1991.
Proceedings of the International Symposium on Runaway Reactions, 1989
CCPS/AIChE Directory of Chemical Process Safety Services

GUIDELINES FOR
SAFE PROCESS
OPERATIONS AND
MAINTENANCE

CENTER FOR CHEMICAL PROCESS SAFETY

of the

AMERICAN INSTITUTE OF CHEMICAL ENGINEERS

345 East 47th Street • New York, NY 10017

Copyright © 1995
American Institute of Chemical Engineers
345 East 47th Street
New York, New York 10017

Library of Congress Cataloging-in Publication Data
Guidelines for safe process operations and maintenance / prepared by
    RMT/Jones and Neuse, Inc.
        p.   cm.
    Includes bibliographic references and index.
    ISBN 0-8169-0627-0 : $130.00
    1. Chemical industry—Safety measures.   I. RMT/Jones and
Neuse. II. American Institute of Chemical Engineers. Center for
Chemical Process Safety.
TP149.G8378   1995
660'. 2804'068—dc20   94–46233
                                                                    CIP

PRINTED IN THE UNITED STATES OF AMERICA
5 4 3 2 1   1 2 3 4 5

This book is available at a special discount when ordered in bulk quantities. For information, contact the Center for Chemical Process Safety of the American Institute of Chemical Engineers at the address shown above.

Contents

8. MAINTENANCE   203

9. SHUTDOWN   245

PREFACE

The American Institute of Chemical Engineers (AIChE) has a 30-year history of involvement with process safety and loss control issues in the chemical, petrochemical, and hydrocarbon process industries. AIChE publications and symposia are information resources for the chemical engineering profession on the causes of process incidents and means of preventing their occurrences or mitigating their consequences.

The Center for Chemical Process Safety (CCPS), a directorate of AIChE, was established in 1985 to develop and disseminate technical information for use in the prevention of major chemical process incidents. With the support and direction of the CCPS Advisory and Managing Boards, a multifaceted program was established to address the need for process safety management systems in industry to reduce potential exposures to the public and the environment. Over 80 corporations from all segments of the process industries provide the funding and professional experience for the Center's activities.

In 1989, CCPS published the *Guidelines for Technical Management of Chemical Process Safety*, which developed a model for a safety management system characterized by twelve distinct and essential elements. The Foreword to that project states: "For the first time, all the essential elements and components of a model of a technical management program in chemical process safety have been assembled in one document. We believe these *Guidelines* provide the umbrella under which all other CCPS Technical Guidelines will be promulgated."

The *Guidelines* mentioned above was followed by a number of other publications which provided detailed guidance on the implementation of each of the elements of the process safety program. The level of detail and description provided in these subsequent publications is intended to provide sufficient guidance for the industry to establish, implement, and practice all elements of the process safety program. However, all these publications still left a need for a guidance document for the front-line operations and maintenance supervisors with regard to their role in the implementation and practice of all the elements of the process safety program in their daily activities. The

Guidelines for Safe Process Operations and Maintenance is intended to fill the need for providing process safety guidance to the first- and second-line supervisors in the conduct of their daily activities through the life cycle of the plant.

This *Guidelines* is organized in ten chapters. The first chapter provides an introduction and a picture of the relevance of this book to other CCPS publications. The second chapter consists of an introductory discussion of the roles of first- and second-line operations and maintenance supervisors through the various phases of the life cycle of a plant. Each of the remaining chapters is dedicated to one of the phases of the life cycle of a plant, that is, design, construction, pre-startup and commissioning, startup, operation, maintenance, shutdown, decommissioning and demolition. Each of the chapters provide "how-to" guidance, tools, and checklists to assist the supervisors in the implementation and practice of process safety program principles. Examples of incidents have been used extensively to emphasize the importance of practicing these principles.

ACKNOWLEDGMENTS

The American Institute of Chemical Engineers and the Center for Chemical Process Safety thanks all of the members of the Operations and Maintenance Subcommittee for their dedicated efforts and technical contributions to the preparation of the Guidelines. CCPS also expresses appreciation to the members of the Technical Steering Committee for their advice and support.

The Chair of the Operations and Maintenance Subcommittee was Leon Ward of Hoechst Celanese Corporation. The subcommittee members were: Paul Besse, Union Carbide Corporation; David Cottle, American Cyanamid Corporation; Frank Davis, Shell Oil Company; Dave Giffin, B.F. Goodrich; and John Lockwood, BP Americas, Inc. Robert A. Schulze was the CCPS staff liaison and was responsible for the overall administration and coordination of the project.

The members of the Operations and Maintenance Subcommittee also wish to thank their employers for providing time to participate in this project and to the many sponsors whose findings made this project possible.

RMT/Jones and Neuse, Inc., Austin, Texas was the contractor for this project. Dr. M. Sam Mannan was RMT/Jones and Neuse's Project Director. The principal authors were: Dr. M. Sam Mannan; Dr. Dwight B. Pfenning; Will Varnado; and Dr. Harry H. West.

CCPS also gratefully acknowledges the comments and suggestions submitted by the following peer reviewers: Myron L. Casada, JBF Associates; Donald J. Connolley, AKZO Nobel Chemicals, Inc.; Thomas Janicik, Solvay Polymers; Harvey Rosenhouse, FMC Corporation; Leslie Scher, W.R. Grace; John Snell, Occidental Chemical Corporation; John Susil, Hoechst Celanese Corporation; and Lester Wittenberg, AIChE/CCPS. Their insight and thoughtful comments helped ensure a balanced perspective for the *Guidelines*. Many other individuals contributed directly or indirectly towards this arduous task. Gary Tinsley, an operations specialist with Hoechst Celanese Corporation, provided important review from the perspective of the intended audience of this *Guidelines*. Others who helped are Harold Bradbury, Kimberly Garrett, Patsy Haynes, Brenda House, Eric Kiihne, Afroza Mannan, Misty Martin, Anne Pfenning, Travis Welborn, Georgelle West and Sharon Wevill.

LIST OF TABLES

LIST OF FIGURES

GLOSSARY

Accident An unplanned event or sequence of events that results in undesirable consequences. An incident with specific safety consequences or impacts.

Autoignition Temperature The autoignition temperature of a substance, whether solid, liquid, or gaseous, is the minimum temperature required to initiate or cause self-sustained combustion, in air, with no other source of ignition.

Autocatalytic Reaction Reaction of which the rate is increased by the catalyzing effect of its reaction products.

Automated System A control system in which programmable electronic systems (PESs) are incorporated by or for the user, but which also contains other components including their application programs.

Availability The fraction of time that the system is actually capable of performing its mission (Dimensionless).

Availability – Uptime/Total Time

Availability – Uptime/(Uptime + Downtime)

The fraction of time a system is fully operational.

Catastrophic A loss of extraordinary magnitude in physical harm to people, with damage and destruction to property, and/or to the environment.

Cause An event, situation, or condition which results, or could result, directly or indirectly in an accident or incident.

Chronic Persistent, prolonged and repeated. Relating to exposure: frequent, or repeated, or continuous exposure to substances. Relating to effects: when physiological effects appear slowly and persist for a long period or with frequent recurrences.

Consequence The cumulative, undesirable result of an incident, usually measured in health/safety effects, environmental impacts, loss of property, and business interruption costs.

Consequence Analysis The analysis of the expected effects of an incident, independent of its likelihood.

Contributing Cause Physical conditions, management practices, etc. that facilitated the occurrence of an incident.

Control Logic A control system in which definite output signal states are functions of the states of the Input signals in keeping with the rules of Boolean algebra.

Cryogenic Of or relating to the production of very low temperatures.

Demand A plant condition or event which requires a protective system or device to take appropriate action in order to prevent a hazard. (1) A signal or action that should change the state of a device, or (2) an opportunity to act, and thus, to fail.

Dense Gas A gas with density exceeding that of air at ambient temperature.

Emergency and First Aid Procedures Actions that should be taken at the time of chemical exposure before trained medical personnel arrive.

Emergency Sequence An automatic sequence initiated by an interlock. The sequence may consist of starting, stopping, opening, or closing equipment in order to render the process safe.

Ergonomics An applied science concerned with designing and arranging things people use so that the people and things interact most efficiently and safely.

Evaporation Rate A number showing how fast a liquid will evaporate. *Importance:* The higher the evaporation rate, the greater the risk of vapors collecting in the workplace. The evaporation rate can be useful in evaluating the health and fire hazards of a material.

Event An occurrence related to equipment performance or human action, or an occurrence external to the system that causes system upset. In this document an event is either the cause of or a contributor to an incident or accident, or is a response to an accident's initiating event.

Event Sequence A specific unplanned sequence of events composed of initiating events and intermediate events that may lead to an incident.

Event Tree Analysis A graphical logic model that identifies and quantifies possible outcomes following an initiating event.

Exothermic A reaction is called exothermic if energy is released during the reaction.

Explosions A release of energy that causes a pressure discontinuity or blast wave.

External Event Event caused by (1) a natural hazard—earthquake, flood, tornado, extreme temperature, lighting, etc.; or (2) man-induced events—aircraft crash, missile, nearby industrial activity, or an interruption of facilities such as electric power or process air.

Fail-safe A concept that defines the failure direction of a component/system as a result of specific malfunctions. The failure direction is toward a safer or less hazardous condition.

Failure An unacceptable difference between expected and observed performance.

Failure Mode A symptom, condition, or fashion in which hardware fails. A mode might be identified as loss of function; premature function (function without demand); an out-of-tolerance condition; or a simple physical characteristic such as a leak (incipient failure mode) observed during inspection.

Failure Mode and Effects Analysis A process for hazard identification where all known failure modes of components or features of a system are considered in turn and undesired outcomes are noted.

Fault Tolerant Refers to a computer program or system where some parts may fail but the system will still execute properly. The control system configuration that inherently provides auto selection of alternate or redundant signal paths to effect uninterrupted operations.

Fault Tree A method for representing the logical combinations of various system states that lead to a particular outcome (top event).

Fault Tree Analysis Estimation of the Hazardous incident. (Top Event) frequency from a logical mode of the failure mechanisms of a system.

Flammability The ability of a material to generate a sufficient concentration of combustible vapors to produce a flame, if ignited.

Frequency The number of occurrences at which observed or predicted events occur per unit time.

Functional check/test Direct test of unit in system, as opposed to off-line mechanical test.

Functional Design A phase in the development of computerized systems which produces detailed descriptions of the system that are independent of particular hardware and software. This includes all flows of information, timing diagrams, and state transition diagrams.

Hard Wired Interlock An interlock accomplished by electro-relays and/or wires. An interlock not accomplished through a programmable electronic system.

Hardware Physical equipment directly involved in performing industrial process measuring and controlling functions.

Hardwired That portion of the logic which is executed by electrical circuits comprised exclusively of hardware.

Hazard A chemical or physical condition that has the potential for causing damage to people, property, the environment, etc.

Hazard Analysis The identification of undesired events that lead to the materialization of a hazard, the analysis of the mechanisms by which these undesired events could occur and usually the estimation of the consequences.

Hazard and Operability Study (HAZOP) To question every part of the process to discover what deviations from the intention of the design can occur and what their causes and consequences may be. This is done systematically by applying suitable guide words. This is a systematic detailed review technique for both batch or continuous plants which can be applied to new or existing processes to identify hazards.

Hazard Zone The zone or region where hazard impact has the potential to occur.

Historic Incident Data Data collected and recorded from past incidents.

Human Error Any human action (or lack thereof) that exceed some limit of acceptability (i.e., an out-of-tolerance action) where the limits of human performance are defined by the system. Includes actions by designers, operators, or managers that may contribute to or result in accidents.

Human Factors A discipline concerned with designing machines, operations, and work environments so that they match human capabilities, limitations, and needs. Includes any technical work (engineering, procedure writing, worker training, worker selection, etc.) related to the human factor in operator–machine systems.

Human–machine interface The operators' windows to monitoring and keys, knobs, switches, etc. for making adjustments in the process.

Human Reliability The study of human errors.

Human Reliability Analysis (HRA) A method used to determine the probability that system-required human-actions, tasks, or jobs will be completed successfully within a required time period. Also used to determine the probability that no extraneous human actions detrimental to the system will be performed.

Incident An unplanned event or series of events and circumstances that may result in an undesirable consequence.

Incident Investigation The management process by which underlying causes of undesirable events are uncovered and steps are taken to prevent similar occurrences.

Incident Investigation Team A group of qualified people that examine an incident in a manner that is timely, objective, systematic, and technically sound to determine that factual information pertaining to the event is documented, probable cause(s) are ascertained, and complete technical understanding of such an event is achieved.

Incident Outcome The physical manifestation of an incident.

Inherently Safe A system is inherently safe if it remains in a nonhazardous situation after the occurrence of nonacceptable deviations from normal operating conditions.

Inhibitor A chemical which is added to another substance to prevent an unwanted chemical change from occurring. *Importance:* Inhibitors are sometimes listed on a MSDS, along with the expected time period before the inhibitor is used up and will no longer prevent unwanted chemical reaction.

Initiating Event The first event in an event sequence. Can result in an accident unless engineered protection systems or human actions intervene to prevent or mitigate the accident.

Injury Physical harm or damage to a person resulting from traumatic contact between the body and an outside agency or exposure to environmental factors.

Interlock (1) A protective response which is initiated by an out-of-limit process condition. (2) Instrument which will not allow one part of a process to function unless another part is functioning. (3) A device such as a switch that prevents a piece of equipment from operating when a hazard exists. (4) To join two parts together in such a way that they remain rigidly attached to each other solely by physical interference. (5) A device to prove the physical state of a required condition, and to furnish that proof to the primary safety control circuit.

Interlock System (1) A system that detects out-of-limits, or abnormal conditions or improper sequences and either halts further action or starts corrective actions. (2) A set of protective instrumentation and controls used to recognize a condition that may become hazardous and which will act to eliminate the condition before it can cause injury to personnel or damage to equipment. See NFPA Standard 85A.

Interlocking A system that detects out-of-limits or abnormal conditions or improper sequences and either halts further action or starts corrective action.

Intermediate Event An event that propagates or mitigates an initiating (basic) event during the accident sequence (e.g., improper operation actions, failure to stop an ammonia leak but an emergency plan mitigates the consequences).

Interrupt A break in the normal flow of a system or program occurring in such a way that the flow can be resumed from that point at a later time.

Interview A cooperative informal meeting with a witness where questions are answered voluntarily.

Lethal Service Service utilizing poisonous gases or liquids of such a nature that a very small amount of the gas or the vapor of the liquid, mixed or unmixed with air, is dangerous to life when inhaled. This class includes substances of this nature that are stored under pressure, or may generate a pressure if stored in a closed vessel (ASME Boiler and Pressure Vessel Code, Section VIII, Div. I).

Likelihood A measure of the expected occurrence of an event. This may be expressed as a frequency (e.g., events/year), a probability of occurrence (given that a precursor event has occurred).

Logic System (1) A group of interconnected logic elements that act in combination to perform a relatively complex logic function. Programming-recording system constructed of solid-state modules based on a series of a binary logic (go/no go) components. (2) A pneumatic or electronic system composed of relays, solid-state electronic system composed of relays, solid-state electronic logic modules, fluidic logic elements, PLCs or DCSs which solves complex problems of interlocking or sequencing through the repeated use of simple functions that define basic concepts such as OR and AND gates.

Medical Treatment As defined by OSHA, treatment (other than first aid) administered by a physician or by registered professional personnel under the standing orders of a physician.

Near-Miss An extraordinary event that could have reasonably resulted in a negative consequence (accident or incident) under slightly different circumstances, but actually did not.

Network A maximal interconnected group of graphical Elements of a Ladder Diagram Program, excluding the left and right Power Rails.

NFPA Acronym for National Fire Protection Association.

NIOSH National Institute for Occupational Safety and Health of the Public Health Service, U.S. Department of Health and Human Services (DHHS). *Importance:* Federal agency which—among other activities—tests and certifies respiratory protective devices, recommends occupational exposure limits for various substances and assists in occupational safety and health investigations and research.

Plant-Specific Data Data which pertain to a unique population of equipment specific to a particular operating plant.

Probability The expression for the likelihood of occurrence of an event or an event sequence during an interval of time or the likelihood of the success or failure of an event on test or on demand. By definition probability must be expressed as a number ranging from 0 to 1.

Process change Any change in process, equipment, material, etc. which requires the implementation of management of change procedures.

Process Hazard Analysis An organized effort to identify and evaluate hazards associated with chemical processes and operations to enable their control. This review normally involves the use of qualitative techniques to identify and assess the significance of hazards. Conclusions and appropriate recommendations are developed. Occasionally, quantitative methods are used to help prioritize risk reduction.

Process Safety A discipline that focuses on the prevention of fires, explosions, and accidental hazardous releases at process facilities. Excludes classic worker health and safety issues involving working surfaces, ladders, protective equipment, etc.

Process Safety Management A program or activity involving the application of management principles and analytical techniques to ensure the safety of process facilities.

Products Chemicals produced during a reaction process.

Program A series of actions proposed in order to achieve a certain result.

Program Library A collection of available computer programs and routines.

Programmable Electronic System (PES) A computer-based system connected to sensors and final control elements for the purpose of control, protection, or monitoring.

Programmable Logic Controller (PLC) (1) A control device, normally used in industrial control applications, that employs the hardware architecture of a computer and (typically) a relay ladder diagram language. (2) Although these are more commonly called Programmable Controllers, the acronym PLC is used in this (book) because PC is more commonly used in referring to a personal computer.

Protective System The collection of hardware and methodologies that comprise the effort to maintain a safe and operable plant in the event of failure in control systems or procedures (e.g., pressure vessel relief valves, that function to prevent or mitigate the occurrence of an incident).

Qualitative Methods Refers to the method of evaluation gained through experience (e.g., application, operation, support) or essential, required and desirable features.

Raw Data The original records from which reliability data is extracted; the facility records of equipment failure, repair, outage, and exposure hours or demands which require analysis and encoding in order to be placed into data elements.

Reactants Chemicals that are converted into the required products during the reaction process.

Reaction The process in which chemicals/materials (reactants) are converted to other chemicals/materials (products). Types of reactions are often named individually e.g.:
oxidations (= oxidations reactions),
decompositions (= decomposition reactions),
bromations (= reactions with Bromine).

Reaction Kinetics The complex of data (thermodynamical and kinetical), that determine a reaction.

Reactive Substance/Material Substance or material which enters into a chemical reaction with other stable or unstable material.

Redundancy The employment of two or more devices, each performing the same function, in order to improve reliability.

Reliability The probability that an item is able to perform a required function under stated conditions for a stated period of time or for a stated demand.

Reliability Analysis The determination of reliability of a process, system, or equipment.

Review (Process Safety Review) An inspection of a plant/process unit, drawings, procedures, emergency plans, and/or management systems, etc., usually by an on-site team and usually problem-solving in nature.

Risk A measure of potential economic loss or human injury in terms of the probability of the loss or injury occurring and the magnitude of the loss or injury if it occurs.

Risk Analysis The development of a quantitative estimate of risk based on engineering evaluation and mathematical techniques for combining estimates of incident consequences and frequencies.

Risk Assessment The process by which the results of a risk analysis (i.e., risk estimates) are used to make decisions, either through relative ranking of risk reduction strategies or through comparison with risk targets.

Risk Contour Lines that connect points of equal risk around the facility ("isorisk" lines).

Risk Estimation Combining the estimated consequences, and likelihood of all incident outcomes from all selected incidents to provide a measure of risk.

Risk Evaluation The assessment of risk, coupled with an appraisal of the significance of the results, both overall and from individual events.

Risk Management The systematic application of management policies, procedures, and practices to the tasks of analyzing, assessing, and controlling risk in order to protect employees, the general public, and the environment as well as company assets, while avoiding business interruptions. Includes decisions to use suitable engineering and administrative controls for reducing risk.

Risk Measures Ways of combining information on likelihood with the magnitude of loss or injury (e.g., risk indexes, individual risk measures, and societal risk measures).

Root Cause(s) A prime reason why an incident occurred. Root causes often are related to deficiencies in management systems.

Safety The expectation that a system does not, under defined conditions, lead to a state in which human life, economics or environment are endangered. *Note:* For system safety, all causes of failures which lead to an unsafe state shall be included; hardware failures, software failures, failures due to electrical interference, due to human interaction and failures in the controlled object. Some of these types of failure, in particular random hardware failures, may be quantified using such measures as the failure rate in the dangerous mode of failure or the probability of the protection system failing to operate on demand. The system safety also depends on many factors which cannot be quantified but can only be considered qualitatively.

Safety-Critical Those systems whose sole function is to maintain the safety of the process, such as a pressure relief valve.

Safety Interlocking Same as interlocking except a failure to control out-of-limit conditions can cause injury or unacceptable environmental contamination.

Serious Injury The classification for an occupational injury which includes all disabling work injuries and nondisabling work injuries as follows: eye injuries requiring treatment by a physician, fractures, injuries requiring hospitalization, loss of consciousness, injuries requiring treatment by a doctor and injuries requiring restriction of motion or work, or assignment to another job.

Software The programs, procedures, and related documentation associated with a system design and system operation. The system can be a computer system or a management system.

Subsystem A portion of a system.

Synchronous Pertaining to two or more processes that depend upon the occurrence of a specific event such as a common timing signal.

Synonym Another name or names by which a material is known. Methyl alcohol, for example, is also known as methanol, or wood alcohol. *Importance:* A MSDS will list common name(s) to help identify specific materials.

System A collection of people, machines and methods organized to accomplish a set of specific functions.

Task Analysis An analytical process for determining the specific behaviors required of the human components in a man–machine system. It involves determining the detailed performance required of people and equipment and the effects of environmental conditions, malfunctions, and other unexpected events on both. Within each task to be performed by people, behavioral steps are analyzed in terms of (i) the sensory signals and related perceptions, (ii) the decisions, memory storage, and other mental processes, and (iii) the required responses.

Toxicity The quality, state, or degree to which a substance is poisonous and/or may chemically produce an injurious or deadly effect upon introduction into a living organism.

Underlying Causes Actual root causes.

Witness A person who has information related, directly or indirectly, to the accident or incident.

1

INTRODUCTION

1.1. PROCESS SAFETY MANAGEMENT ACTIVITIES OF THE CENTER FOR CHEMICAL PROCESS SAFETY (CCPS)

This book, *Guidelines for Safe Process Operations and Maintenance*, was prepared to offer practical guidance to first- and second-level supervisors and managers in executing a process safety management program. It includes examples of process safety management programs for facilities that manufacture, handle, or use hazardous chemicals. The *Guidelines* are part of a CCPS multifaceted program whose purpose is to enhance chemical process safety. The CCPS program sponsors the following publications and activities:

- An overview brochure entitled "A Challenge to Commitment"[1] is addressed to the chief executive officers of the 1500 member companies of the chemical process industry. This brochure emphasizes that it is essential that the chemical process safety program have top management's understanding and support to reduce accidents and improve the safety record of the chemical process industry. The brochure briefly describes the 12 technical elements of a model chemical process safety program (see Table 1-1).
- The flagship publication of the CCPS program is *Guidelines for the Technical Management of Chemical Process Safety*[2]. This publication expands on the 12 elements described in the overview brochure and targets middle management. It describes the framework and components of the CCPS chemical process safety management system (see Table 1-2). A subsequent publication, *Plant Guidelines for Technical Management of Chemical Process Safety*,[3] is more detailed to facilitate the implementation of process safety management programs at the plant level. The *Plant Guidelines* book is supplemented by a training course to aid the understanding and application of the *Guidelines* and present examples of plant programs.

1

TABLE 1-1
The Twelve Elements of Chemical Process Safety Management

 1. Accountability: Objectives and Goals

 2. Process Knowledge and Documentation

 3. Capital Project Review and Design Procedures (for new and existing plants, expansions, and acquisitions)

 4. Process Risk Management

 5. Management of Change

 6. Process and Equipment Integrity

 7. Human Factors

 8. Training and Performance

 9. Incident Investigation

10. Standards, Codes, and Laws

11. Audits and Corrective Actions

12. Enhancement of Process Safety Knowledge

- *Guidelines for Safe Process Operations and Maintenance* is part of a series that develops sections of the program in greater detail. The focus of the *Guidelines* is on the *execution* of policies and procedures for safe process operations and maintenance, rather than on the formulation of these policies. Figure 1-1 illustrates the relationship of the *Guidelines* book to other CCPS publications and activities.
- CCPS is actively involved in identifying gaps in technology and process safety management systems and in developing research projects to fill these gaps. In addition, CCPS conducts international symposia and workshops to encourage a worldwide exchange of information on the subject.

1.2. PROCESS SAFETY ACTIVITIES OF GOVERNMENTAL AGENCIES AND TRADE ORGANIZATIONS

During the past 15 years, a number of chemical or related incidents in the petrochemical industry have adversely affected surrounding communities. A few of these incidents, such as the vapor cloud explosion in Flixborough in 1974, the LPG explosion in Mexico City in 1984, the toxic material release in Bhopal in 1984, and the fire and radiation release in Chernobyl, were reported worldwide. Both governmental agencies and trade organizations responded by developing standards and regulations to improve process safety. The American Petroleum Institute (API) and the Chemical Manufacturers Association (CMA) started to work with their members to develop organizational guidelines. The U.S. Department of Labor directed the Occupational Safety

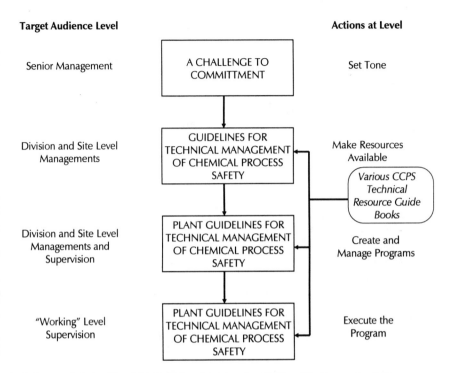

FIGURE 1-1. Relationship of this *Guidelines* book to other CCPS publications and activities.

and Hazard Administration (OSHA) to develop federal standards for managing process safety.

A consensus started to emerge in 1990. Although the language, application, and extent of each document differed, the contents and objectives were almost the same. The API published *Recommended Practice 750: Management of Process Hazards*[4] in January 1990. OSHA published the proposed federal process safety rule[5] in July 1990. In October 1990, the CMA published its *Resource Guide for Implementing the Process Safety Management Code of Practices*[6]. In addition, the *Clean Air Act Amendments of 1990* directed OSHA and the Environmental Protection Agency (EPA) to develop process safety management regulations to protect workers and the environment. The final OSHA rule on Process Safety Management of Hazardous Chemicals (29 CFR 1910.119) was published in the *Federal Register*[7] on February 24, 1992. Appendix A contains a summary of the requirements of the OSHA rule. A matrix showing the relevance of OSHA PSM elements to CCPS's chemical process safety management elements is also given in Appendix A.

The international chemical and petroleum community has also been addressing process safety management through regulations and recom-

TABLE 1-2
Elements and Components of Process Safety Management

1. **Accountability: Objectives and Goals**
   Continuity of Operations
   Continuity of Systems (resources and funding)
   Continuity of Organizations
   Company Expectations (vision or master plan)
   Quality Process
   Control of Exception
   Alternative Methods (performance vs.
     specification)
   Management Accessibility
   Communications

2. **Process Knowledge and Documentation**
   Process Definition and Design Criteria
   Process and Equipment Design
   Company Memory (management
     information)
   Documentation of Risk Management
     Decisions
   Protective Systems
   Normal and Upset Conditions
   Chemical and Occupational Health Hazards

3. **Capital Project Review and Design
   Procedures (for new or existing plants,
   expansions, and acquisitions)**
   Appropriation Request Procedures
   Risk Assessment for Investment Purposes
   Hazards Review (including worst credible cases)
   Siting (relative to risk management)
   Plot Plan
   Process Design and Review Procedures
   Project Management Procedures

4. **Process Risk Management**
   Hazard Identification
   Risk Assessment of Existing Operations
   Reduction of Risk
   Residual Risk Management (in-plant
     emergency response and mitigation)
   Process Management during Emergencies
   Encouraging Client and Supplier Companies
     to Adopt Similar Risk Management Practices
   Selection of Businesses with Acceptable Risks

5. **Management of Change**
   Change of Technology
   Change of Facility
   Organizational Changes That May Affect
     Process Safety
   Variance Procedures
   Temporary Changes
   Permanent Changes

6. **Process and Equipment Integrity**
   Reliability Engineering
   Materials of Construction
   Fabrication and Inspection Procedures
   Installation Procedures
   Preventive Maintenance
   Process, Hardware, and Systems Inspections
     and Testing (pre-startup safety review)
   Maintenance Procedures
   Alarm and Instrument Management
   Demolition Procedures

7. **Human Factors**
   Human Error Assessment
   Operator/Process and Equipment Interfaces
   Administrative Controls versus Hardware

8. **Training and Performance**
   Definition of Skills and Knowledge
   Training Programs (e.g., new employees,
     contractors, technical employees)
   Design of Operating and Maintenance
     Procedures
   Initial Qualification Assessment
   Ongoing Performance and Refresher Training
   Instructor Program
   Records Management

TABLE 1-2 (Continued) Elements and Components of Process Safety Management	
9. Incident Investigation Major Incidents Near-miss Reporting Follow-up and Resolution Communication Incident Recording Third-party Participation as Needed	**11. Audits and Corrective Actions** Process Safety Audits and Compliance Reviews Resolutions and Close-out Procedures
10. Standards, Codes, and Laws Internal Standards, Guidelines, and Practices (past history, flexible performance standards, amendments, and upgrades) External Standards, Guidelines, and Practices	**12. Enhancement of Process Safety Knowledge** Internal and External Research Improved Predictive Systems Process Safety Reference Library

mended practices. The Norwegian Petroleum Directorate issued rules[8] in 1981 requiring quantitative hazard analyses for offshore petroleum operations. In response to the 1976 chemical dioxin release in Seveso, Italy, a European Directive[9] (commonly called the Seveso Directive) on process safety management was issued in 1982. More recently, the British government has issued process safety management regulations[10] for North Sea petroleum operations, following the recommendations of the widely distributed Cullen Report, which investigated the 1985 Piper Alpha offshore platform tragedy. Outside of Europe, the World Bank[11] has provided process safety management guidance for third-world projects. Similarly, the International Labor Office in Geneva has issued hazard analysis recommendations[12].

Thus, in summary, the process safety management standards promulgated by OSHA are in stride with the recent activities of international organizations. This *Guidelines* book, along with other CCPS publications referred to in Figure 1-1, gives the necessary guidance for developing and implementing the process safety management programs required by the OSHA standards.

1.3 TARGET AUDIENCE AND OBJECTIVE OF THIS DOCUMENT

This book is written for the plant personnel who must execute site safety programs, policies, and procedures during the life cycle of a plant, starting with initial design and continuing through the entire life cycle of the plant. The primary target audience includes all operations and maintenance personnel who have first- and second-line supervision responsibilities. These are the people who must deal first hand with processes containing hazardous materials and who directly supervise operations and maintenance personnel. Thus, they are in a position to be able to directly affect process safety. Since the nature of operations and maintenance activities differs at each facility, the application of the procedures described in this book will vary. Each company should structure its process safety management system to its culture, its mission, and its business.

The importance of the first-line supervisor in the overall safety of a process plant is extremely critical since the supervisor is the key person who manages personnel, machines, and working conditions on a daily basis. The first-line supervisor is the management person closest to the operating and maintenance personnel who are most likely to get hurt by a process accident; this supervisor is therefore a very important link in the organizational safety chain. The first-line supervisor is also in a position to take early actions to prevent minor circumstances from escalating into an undesirable event.

All the elements of a process safety program, both oversight and actual implementation, fall to some degree into the province of the operations and maintenance departments. Some activities will be supportive in nature, whereas other functions may require a high degree of primary responsibility. For example, conducting operations according to established written procedures is a constant primary activity; training is an example of a process safety support activity that helps operations and maintenance personnel with their tasks. The intent of this document is to describe the recommended technical and administrative process safety activities for operations and maintenance personnel.

The objective of this *Guidelines* then, is to define the process safety responsibilities and roles of operating and maintenance personnel and their direct supervisors and provide guidance for carrying out those responsibilities and roles. The audience is the personnel at a plant site who deal first hand with hazardous materials and who are responsible for managing and controlling real-life process hazards. These personnel are especially susceptible to the pressures of production demand and sensitive to situations that affect human error and reliability. Those responsible for maintaining the integrity of facilities, controlling the manufacturing process, and handling unusual or emergency conditions are the target audience of this document.

The primary target audience is not persons who create site safety policy but those who develop plant-level programs to conform to the defined policy and guidelines. Organizationally this means, the first- and second-line supervisors, who may have titles such as "foreman," "supervisor," or "superintendent."

In a typical plant organization, the plant manager and his staff will define the plant policies and procedures that govern safe operations and maintenance. These policies and procedures are executed at the operating department level. An operating department is a distinct process unit that may have a department superintendent who may be an engineer with overall responsibility for the operation and maintenance of the unit. Most operating unit organizations have a day operations supervisor reporting to the department superintendent. Reporting to the day operations supervisor may be several shift supervisors responsible for directly supervising the operators running the unit. Depending on the philosophy of the company, the positions of the day operations supervisor and shift supervisor may be filled either by individuals who have engineering or other technical degrees or by individuals with many years of practical, hands-on experience as operators.

Plant organizations vary, depending on the culture, business, and nature of the operations and maintenance activities in different plants. However, an example of a simple organizational structure is shown in Figure 1-2 and can be used to understand the type of personnel for whom this document is written. Plant maintenance organizations can be aligned with operating units and have a similar supervisory structure, as illustrated in the example in Figure 1-2.

The *Guidelines* illustrate where and how the contributing process safety roles of the target audience of this document mesh with the overall matrix of roles in a process operation. This matrix begins with the tone set by senior

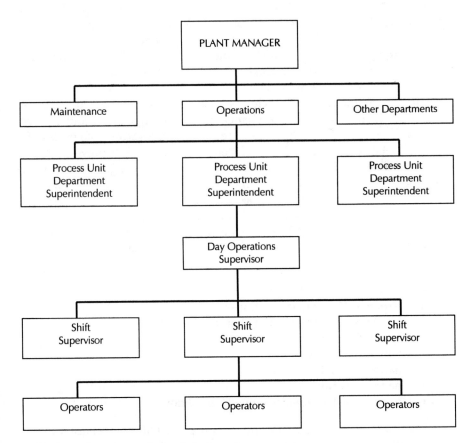

FIGURE 1-2. Example organizational structure.

management. The *Guidelines* illustrate the hand-over roles during the various life cycles of a plant; for example, from the design and construction teams to the startup committee, or from operations to maintenance during a turn-around.

1.4. USE OF THIS DOCUMENT

The organization of this book by life cycle stages allows for easy access to information. It can be used to reference what can be done early in the life of a new plant to provide for future needs. Introducing some or all of these plant safety concepts into the culture of an existing operation will be facilitated by referring to the appropriate life cycle stage. While the *Guidelines for the Technical Management of Chemical Process Safety*[2] and the *Plant Guidelines for Technical Management of Chemical Process Safety*[3] contain policy guidance, this book offers practical, "how-to" reference on process safety management program execution in the operations and maintenance departments.

Managing change during an operation's life is a key issue and has been given major attention in this *Guidelines* document because operations and maintenance personnel ultimately initiate, implement, and execute the changes. Throughout the life cycle of a plant, continuing safety depends on the proper management and implementation of changes in facilities, processes, and utilities. All changes proposed during the various life cycles of a plant must be evaluated for their effect on safety, health, loss prevention, and the environment before they are implemented. Since management of change has a key role in safe process operations and maintenance, the subject is discussed in detail in Chapter 2, Section 2.12. In addition, the relevance of managing change during each phase of the life cycle of the plant is discussed in the appropriate chapter. The elements of process safety that relate to each stage of project life (listed below) are contained in the chapter describing that project life cycle stage.

- Chapter 3—Plant Design
- Chapter 4—Plant Construction
- Chapter 5—Pre-startup and Commissioning
- Chapter 6—Startup
- Chapter 7—Operation
- Chapter 8—Maintenance
- Chapter 9—Shutdown
- Chapter 10—Decommissioning/Demolition

The reader is directed to the appropriate project life cycle chapter to find the specific process safety management elements of concern.

1.5. REFERENCES

1. Center for Chemical Process Safety. "A Challenge to Commitment." American Institute of Chemical Engineers, New York, 1985.
2. Center for Chemical Process Safety. *Guidelines for Technical Management of Chemical Process Safety*, American Institute of Chemical Engineers, New York, 1989.
3. Center for Chemical Process Safety. *Plant Guidelines for Technical Management of Chemical Process Safety.* American Institute of Chemical Engineers, New York, 1992.
4. American Petroleum Institute. *Recommended Practice 750: Management of Process Hazards*. API, Washington, D.C., 1990.
5. *Federal Register.* "Notice of proposed rule making on Process Safety Management of Highly Hazardous Chemicals": 29 CFR 1910.119. Washington, D.C., July 17, 1990.
6. Chemical Manufacturers Association. *Resource Guide for Implementing the Process Safety Management Code of Practices.* Washington, D.C., 1990.
7. Federal Register. "Final rule on Process Safety Management of Highly Hazardous Chemicals": 29 CFR 1910.119, Washington, D.C., February 24, 1992.
8. Norwegian Petroleum Directorate. "Safety Evaluation of Platform Conceptual Design." Stavanger, Norway, 1981.
9. European Community Directive. "On the Major Accident Hazards of Certain Industrial Activities." 82/501/EC. Journal of the European Community. L230. June 1982.
10. *Offshore Installation (Safety Case) Regulation 1992*, Health and Safety Executive, London, U.K.
11. World Bank Technical Paper #55. "Techniques for Assessing Industrial Hazards." Washington, D.C., 1988.
12. *Major Hazard Control*, a practical manual, International Labour Office, Geneva, Switzerland, 1988.

2

ROLE OF OPERATIONS AND MAINTENANCE IN PROCESS SAFETY MANAGEMENT

The operations and maintenance departments are at the core of the process safety management program in any facility. These departments are essential to implementing all the elements of the program; however, the nature and extent of their involvement with each element varies. The remaining chapters of this book are devoted to the process safety roles of the operations and maintenance department first- and second-line supervisors in each particular phase in the life cycle of the plant. In contrast, this chapter describes the role of the operations and maintenance departments in executing each process safety element (listed in Table 1-2) during the various phases of the life cycle of the plant.

Managing change is a key process safety element, and operations and maintenance supervisors are the only ones who can ensure that management of change principles and procedures are implemented properly. A large part of this chapter, therefore, is devoted to the role of these supervisors in executing management of change activities. Their level of involvement, as shown below, depends on the nature of the activity. These role definitions are only included as examples and may vary from company to company depending on their PSM organizational structure.

- **Lead role**: This is the highest level of involvement and control. A person in the lead role usually initiates the necessary activity and may also maintain control and direction over it. An example is the lead role of the first- and second-line supervisors in measuring the on-the-job safety performance of an operations and maintenance employee.
- **Participatory role**: The involvement at this level is high but the level of control may vary from low to high, depending on the nature of the activity. An activity may be initiated by a person having a participatory

role, but this person may not control or direct the activity. For example, although operations and maintenance personnel may initiate the request for a what-if analysis, and they may also participate in the study, they typically do not control or direct the study. In this case, the operations and maintenance role is defined as participatory.

- **Supportive role**: This role is defined for situations where operations and maintenance personnel act as a resource to others and participate when requested. For example, plant management may identify a need for a fault tree analysis to determine the probability of a certain event occurring. Operations and maintenance personnel may provide all the data needed by the analysis team; however, these personnel do not play any role in the actual analysis. Thus, they provide only support and assistance for the activity.

- **Minimal involvement**: This role defines situations where operations and maintenance staff are only minimally involved in the initiation, control, or direction of an activity. For example, operations and maintenance personnel will have only a minimal role in evaluating potential financial risks during capital projects review.

2.1. ACCOUNTABILITY

Accountability is the employee's obligation to explain and answer for those actions that relate to company expectations, objectives, and goals[1,2]. Operations and maintenance staff should help management develop procedures to ensure accountability during the various phases of the life cycle of the plant. For example, a company may have specific guidelines to use in developing shutdown procedures. Operations and maintenance staff participation in this process is threefold. First, they should review and agree with the specific guidelines adopted by management. If the guidelines are not specific, they should be clarified before procedures are developed. Second, once the procedures are in place, the operations and maintenance supervisors are responsible for seeing that they are implemented as stated. If the procedures need to be changed, proper authorization should be obtained. Third, the experience gained from implementing procedures, as well as from any deviations, should form the basis of recommendations to management for formally changing the guidelines or procedures.

Management is responsible for devising a process safety management plan that clearly defines the responsibility and accountability of all functions. Operations and maintenance staff should provide input during the whole process. A process safety management plan is not static. It is changed and updated throughout the life of the plant. Operations and maintenance supervisors may be called on for advice and information when the plan is written and revised. They should be prepared to provide information on both successful and unworkable policies and practices. Their knowledge of everyday plant operation and maintenance activities is crucial to making the plan both effective and efficient.

2.2. PROCESS KNOWLEDGE AND DOCUMENTATION

Capturing and retaining process knowledge is an important element of the process safety management program. Also, regulatory agencies require the documentation of process information and procedures.[3] Operations and maintenance staff can contribute as well as benefit from accurate, complete, and up-to-date process documentation. The involvement of operations and maintenance staff in developing, maintaining, and using documented process knowledge continues through each phase of the life cycle of the plant. During design, operations and maintenance personnel provide input to the project team about specific design criteria and other considerations to be documented. In this way, their process knowledge and experience are incorporated in the design of the plant. In the construction stage, they should be part of the quality assurance team. Their expertise and knowledge should be used to ensure (a) that process equipment is fabricated in accordance with design specifications and that the equipment is assembled and installed correctly, and (b) that the documentation on each item of equipment or process is preserved for future reference. During design and construction, maintenance staff collect information to develop the required maintenance programs. During pre-startup and commissioning, startup, operation, and decommissioning/demolition the operations department has a leadership role in maintaining process knowledge and documentation, and the maintenance departments support them. The roles reverse during the maintenance phase, with maintenance staff taking the leadership role and operations staff taking the participatory role. The operations department retains its leadership role in knowledge of chemical and occupational health hazards even during the maintenance phase of the life cycle of the facility.

2.3. CAPITAL PROJECT REVIEW AND DESIGN PROCEDURES

For this element of process safety management, operations and maintenance representatives have both supportive and participatory roles in reviewing hazard studies, selecting plot plans, reviewing process designs, and developing procedures. In the design stage, their input is both supportive and participatory. They may be required to contribute their knowledge and experience to the project team or to answer specific questions about the operations and maintenance aspects of the plant. They should also participate in the hazard reviews conducted during the design phase.

The capital project review may involve either a new project, major capacity additions, or the manufacture of new products. Capital projects may also include substantial facility modifications. Such projects may involve many types of changes, such as the introduction of new equipment, chemistries, controls, operating procedures, or chemical inventories. The nature and extent of the participation of the operations and maintenance departments varies, depending on the nature of the project and type of analysis required. Table 2-1 shows example activities and the relative involvement of the operations and maintenance departments.

2.4. PROCESS RISK MANAGEMENT

The initial identification and evaluation of process risks from fires, explosions, toxic releases, runaway reactions, or natural disasters is performed during preliminary process design. The operations and maintenance departments participate in these risk reviews as members of a review team, such as a HAZOP team. The analysis includes identifying hazards, evaluating the risks of existing operations, reducing risk, and residual risk management. During pre-startup and commissioning, startup, operation, shutdown, and decommissioning/demolition, the operations department has the leadership role in performing process risk analyses; maintenance plays a supportive role. During the maintenance phase of the life cycle of a plant, the risk analysis is performed under the leadership of the maintenance department. Table 2-2 shows the various phases of the life cycle of a plant, the different types of process risk management techniques employed in each phase of the life cycle, and the extent of operations and maintenance department involvement.

Generally, first- and second-line operations and maintenance supervisors play no role in encouraging client and supplier companies to adopt similar risk management practices. The selection of businesses with acceptable risks is a decision made by management. After design and construction, the only input needed from the operations and maintenance departments are the risk assessment analyses performed after construction is complete.

2.5. PROCESS AND EQUIPMENT INTEGRITY

The continuing integrity of processes and equipment can only be ensured by the operations and maintenance departments. First- and second-line operations and maintenance supervisors play a vital role in reducing or eliminating process upsets or equipment failures that would lead to releases of hazardous materials. They should stay involved with the process and equipment integrity program throughout the life of the plant. However, operations personnel play the lead role in matters of process integrity, while maintenance personnel

TABLE 2-1
Examples of Manufacturing Departments' Roles in Reviewing Capital Project

Description of Capital Project Review Activity	Operations Involvement	Maintenance Involvement
Is the company prepared to accept the potential risks of this project?	Minimal	Minimal
Is the location appropriate for the types of operations envisioned?	Minimal	Minimal
Is there an appropriate buffer zone between the operation and the surrounding facilities or neighborhoods?	Minimal	Minimal
Is there adequate transportation into the site?	Minimal	Minimal
Has the chemistry been thoroughly investigated so that all process hazards are known?	Minimal	Minimal
Does the design adequately address all the process hazards?	Supportive	Supportive
Have the equipment, piping, and control instrumentation been checked for inadvertently designed-in" hazards?	Participatory	Participatory
Does the installed equipment conform to the design, as delineated by the drawings and specifications?	Participatory	Participatory
Have computer and control systems been adequately checked and documented?	Supportive	Participatory
Are the consequences of process deviations known, and are the consequences acceptable? If not acceptable, have mitigating systems or procedures been provided?	Supportive	Supportive
Does the design address the possibility of human failure, and does it adequately handle the consequences of such failure?	Participatory	Participatory
Is there an adequate supply of necessary utilities for commissioning, startup, operations, shutdown, and decontamination?	Participatory	Participatory
Is there a suitable emergency response plan with resources identified or provided?	Participatory	Supportive
How does the estimated process risk compare with that of other existing or proposed operations?	Minimal	Minimal
How are the risks from this process going to be controlled?	Participatory	Participatory
What actions, equipment, procedures, or training are needed to control the risks identified?	Participatory	Supportive

play a supportive role. In contrast, the roles are reversed with regard to equipment integrity; i.e., maintenance staff take the lead and operations staff provide support. Table 2-3 shows the types of input and support provided by the operations and maintenance departments during the different phases of the life cycle of a plant.

TABLE 2-2
Examples of Operations and Maintenance Roles in Process Risk Management during the Different Phases of the Life Cycle of a Plant[a]

Life Cycle Phase	Process Risk Management Activity	Operations Involvement	Maintenance Involvement
Design	Comparative site evaluation	Minimal	Minimal
	Design review	Participatory	Participatory
	Inherent safety analysis	Supportive	Minimal
	Process hazards analysis (i.e., HAZOP study)	Participatory	Participatory
	Quantitative risk assessment	Supportive	Supportive
	Consequence analysis	Supportive	Supportive
Construction	Quality assurance	Supportive	Participatory
	Design verification	Supportive	Participatory
	Checklists	Participatory	Participatory
	Resolution of previous hazards	Supportive	Supportive
Pre-Startup & Commissioning	Startup & operating procedures	Lead	Participatory
	Pre-startup safety review	Lead	Participatory
Startup	Identification of error-likely situations	Lead	Participatory
Operation	Procedures	Lead	Supportive
	Revise HAZOP study	Participatory	Participatory
	Quantitative risk assessment	Supportive	Supportive
Maintenance	Maintenance procedures	Supportive	Lead
	Failure modes and effects	Supportive	Participatory
Shutdown	Procedures	Lead	Supportive
	Pre-shutdown checklist	Lead	Supportive
	Decontamination review	Lead	Participatory
	Equipment checklist	Participatory	Lead
	Process hazards analysis	Participatory	Participatory
Decommissioning	Procedures	Supportive	Participatory
	Mothball checklist	Supportive	Participatory
	Hazard review of decommissioned plant status	Participatory	Participatory

[a]The levels of involvement shown here will vary depending on the company culture and complexity of the process. For specific cases, conscious decisions should be made using site-specific considerations.

2.6. HUMAN FACTORS

Human factors and ergonomics are critical to process safety. First- and second-line supervisors can, by virtue of their position, reduce the number of incidents caused by human factors. During the design and construction phase of a project, operations and maintenance should provide input to the project

TABLE 2-3
Examples of Operations and Maintenance Roles in the Process and
Equipment Integrity Program

Life Cycle Phase	Process and Equipment Integrity Activity	Operations Involvement	Maintenance Involvement
Design	Selection of materials of construction	Minimal	Minimal
	Reliability engineering	Minimal	Minimal
	Fabrication procedures	Minimal	Supportive
	Selection of standards for testing and frequency of testing	Minimal	Supportive
Construction	Installation procedures	Minimal	Supportive
	Create baseline equipment integrity data	Minimal	Supportive
Pre-Startup & Commissioning	Process, hardware, and systems inspection and testing	Minimal	Supportive
	Pre-startup preparation	Participatory	Supportive
Startup	Sequence description	Participatory	Supportive
	Procedures	Lead	Supportive
	Post-startup review	Participatory	Participatory
Operation	Alarm and instrument management	Participatory	Participatory
	Reliability engineering	Supportive	Participatory
	Testing and inspection procedures	Participatory	Participatory
	Management of change	Participatory	Participatory
Maintenance	Maintenance procedures	Supportive	Lead
	Preventive maintenance	Supportive	Participatory
	Predictive maintenance	Minimal	Supportive
	Reliability engineering	Minimal	Supportive
	Testing and inspection of equipment	Supportive	Participatory
	Update equipment integrity data	Minimal	Participatory
	Management of change	Participatory	Participatory
Shutdown	Shutdown procedures	Lead	Supportive
	Pre-shutdown preparation	Lead	Supportive
	Post-shutdown review	Participatory	Participatory
Decommissioning	Demolition procedures	Minimal	Participatory
	Pre-demolition preparation	Minimal	Lead
	Post-demolition review	Minimal	Lead

team about human factors so that the best possible ergonomic design can be achieved and the plant design is less susceptible to human error. During all other phases of the life cycle of the plant, first- and second-line supervisors should use administrative controls, training, and awareness programs to reduce the adverse effects of human errors. Operations and maintenance personnel may play a supportive role in evaluating human error; however, implementing the recommendations drawn from human error evaluation is the primary responsibility of first- and second-line operations and maintenance supervisors.

The role of the operations and maintenance departments in determining the most effective ergonomic design can best be understood by considering an incident that occurred in an ethylene plant. The emergency shut down (ESD) switches on the control panel for shutting down the cracking furnaces were so arranged that the numbering sequence was very confusing. An emergency occurred when a radiant tube for one of the furnaces (furnace #16) ruptured. When the emergency call was received at the control room, the operator shut down the wrong furnace (furnace #17 instead of furnace #16). The error was not detected for a few more minutes, which caused an even more severe emergency and the loss of equipment and materials. A combination of a confusing configuration of ESD switches and operator error during an emergency response increased the severity of the incident. This could have been avoided by making proper ergonomic decisions during design and construction. The first- and second-line operations and maintenance supervisors are in the best position to ensure that such matters are considered by the project team.

2.7. TRAINING AND PERFORMANCE

Site-specific training programs are an essential element of the process safety management program of every facility. The activities associated with training and with the evaluation of performance occur consistently throughout all the life cycle phases of a processing facility. Although the training requirements and training guidelines may be specified or developed by corporate management, the actual training may be conducted by in-house experts or consultants. However, first- and second-line operations and maintenance supervisors play a key role in (a) ensuring that the training programs are designed specifically for the various functions or jobs in a plant, (b) ensuring that all personnel assigned to specific duties have received the proper training, as stated in company guidelines, and (c) communicating any changes and lessons learned from process safety incidents and day-to-day activities to management for use in updating training programs and requirements. Table 2-4 shows the different components of training and performance and the roles of operations and maintenance personnel in accomplishing the activities associated with each component. The first- and second-line supervisors are in the best position to enforce the complete and proper execution of the operating and maintenance procedures. The need to monitor the execution of operating and maintenance procedures is minimal when well-trained, knowledgeable, motivated, and alert staff are working. But the need for oversight is always there. Without vigilance, personnel can drift from the complete and proper execution of procedures, opening the door to a process safety incident.

TABLE 2-4
Examples of the Role of Operations and Maintenance in Implementing
the Different Components of Training and Performance

Training & Performance Component	Description of Activity	Operations Involvement	Maintenance Involvement
Definitions of skills and knowledge	Typically, corporate or plant management establishes guidelines. However, operations and maintenance should make sure that their concerns are known to management.	Participatory	Participatory
Design of operating and maintenance procedures	Operating and maintenance procedures are typically established during design by the project team. However, operations and maintenance should provide their input to the project team and ensure that the procedures are updated and revised as needed.	Participatory	Participatory
Initial qualifications evaluation	Operations and maintenance should provide their input on the initial training requirements for all employees.	Participatory	Participatory
Selection and development of training program	Plant management fulfills this responsibility. Operations and maintenance should ensure that the training objectives are accomplished by the specified training program.	Participatory	Participatory
Measuring performance & effectiveness	Operations and maintenance first- and second-line supervisors should use on-the-job observation of task performance to measure performance and the effectiveness of training.	Lead	Lead
Instructor program	Although corporate and/or plant management may make initial instructor selection, operations and maintenance should provide their observations about each instructor. These observations should be based on the measurement of performance before and after training.	Participatory	Participatory
Records management	Documentation of training and performance should be maintained and managed by operations and maintenance staff and copies should be furnished to others as necessary.	Lead	Lead
Ongoing performance and refresher training	Corporate and/or plant management provides guidelines. However, implementation of those guidelines is the responsibility of operations and maintenance.	Lead	Lead

2.8. INCIDENT INVESTIGATION

Incidents are broadly classified as unplanned events with undesirable consequences[1]. Even under the best of circumstances, when due diligence is used to develop and implement the process safety management program, process safety incidents can occur. These incidents may happen during any phase of the life cycle of the plant except during design. The investigation of incidents should be the vehicle for reducing or eliminating the causes that led to the particular incident and, ultimately, for improving the overall safety of the plant. Incident investigation guidelines and procedures are typically developed by corporate management; however, first- and second-line supervisors should play an active role in developing and implementing the incident investigation procedures. In addition, they should ensure that the recommendations resulting from incident investigations are implemented as soon as possible. Table 2-5 provides examples of incident investigation activities and the extent of operations and maintenance personnel involvement in those activities.

2.9. STANDARDS, CODES, AND REGULATIONS

Plants are required to follow certain governmental regulations, industry standards, and internal company guidelines. The requirements may apply to layout, design, equipment, technology, and training, or to some other facet of day-to-day plant activities. First- and second-line supervisors usually have minimal participation in the adoption and implementation of standards related to layout, design, construction, and equipment. In contrast, first- and second-line operations and maintenance supervisors should actively participate in implementing standards that relate to operating and maintenance procedures and training. For example, consider the case of hot work procedures. Corporate staff may set broad guidelines covering hot work procedures and site management may develop site-specific implementation procedures. However, first- and second-line supervisors are the ones who implement the procedures, issue hot work permits on a day-to-day basis, maintain records, and provide feedback to management about the success or difficulty of implementing and practicing of the procedures.

2.10. AUDITS AND CORRECTIVE ACTION

Process safety audits are required by government regulations and company guidelines. The intent of these audits is to (a) determine the status of compliance with regulations and company guidelines, (b) determine the status and

TABLE 2-5
Examples of the Role of Operations and Maintenance in
Incident Investigation Activities

Incident Investigation Activities	Operations Involvement	Maintenance Involvement
Development of guidelines and procedures	Supportive	Supportive
Implementation of site policies	Participatory	Participatory
Incident investigation		
Fact finding	Participatory	Participatory
Collection of evidence	Participatory	Participatory
Evidence analysis and cause determination	Supportive	Supportive
Documentation	Supportive	Supportive
Follow-up and implementation of recommendations	Participatory	Participatory

cost-effectiveness of safety management efforts, and (c) provide recommendations to management about how to increase compliance and improve the overall safety of the plant. The role of operations and maintenance personnel during audits usually consists of providing information to the audit team. In addition to reviewing records and interviewing plant management, the audit team will typically interview a number of personnel from the operations and maintenance departments during the audit. The audits can therefore be used by operations and maintenance personnel in their continuing quest to improve the process safety management program for the facility. This can best be done by:

- Providing all the data and documentation requested by the audit team, and
- Providing frank and straightforward answers to queries made by the audit team. Employees' concerns and recommendations can thus be incorporated in the process safety management program.

2.11. ENHANCEMENT OF PROCESS SAFETY KNOWLEDGE

Process safety management programs should be dynamic and adaptable to changes in operations, process activities, and personnel. Changes and improvements should be incorporated in the program when needed. First- and second-line supervisors should remain aware of the ongoing changes and enhancements in the process safety management program. Operations and maintenance personnel are the first to know when something is not working;

therefore, they are a key source of information. When new guidelines are adopted or existing guidelines are modified, operations and maintenance personnel should help implement them. They should at least help evaluate the technology gaps and training needs.

2.12. MANAGEMENT OF CHANGE

First- and second-line supervisors of the operations and maintenance department have a key role regarding the recognition of management of change situations and ultimately the execution of proper management of change procedures. Management of change procedures will usually include some or all of the following:

- Local definition of process changes
- The process and mechanical design basis for the proposed change
- Analyses of the safety, health, and environmental considerations
- The effects on separate but interrelated upstream or downstream facilities
- Necessary updates to the operating procedures
- Required training of appropriate personnel
- Necessary updates of process documentation
- Duration of the change
- Required authorization

Operations and maintenance understanding of and input about each of these factors is essential to successfully execute the changes. Another key factor in a successful management of change program is open communication among all parties about the change authorization procedure. All relevant personnel must be informed of the change; documentation of the training is the responsibility of operations and maintenance supervisors.

The development and effective implementation of a change control program is critically dependent on periodic auditing and verification. The audit and verification process associated with the management of change process focuses on the accuracy and completeness of documentation of changes to process facilities. Random checks of a system or subsystem can be completed to determine whether all the changes have been reviewed for their effect on the safety of chemical processes, and whether all changes in hardware, software, and documentation are reflected in the process safety information documentation and database.

2.12.1. Importance of Changes

No chemical process operates completely the same day after day, year after year. Changes are always necessary, sometimes to respond to different economic conditions or sometimes to new technology. Change, however, also presents an opportunity for new hazards or risks to be introduced to the chemical process plant if it is not managed properly. Many accidents can be directly attributed to the unforeseen aspects of a process modification.

Operations and maintenance personnel are in a good position to help manage the safety consequences of changes. One of the most important roles of operations and maintenance personnel is their ability to recognize situations that require management of change procedures. As a group that has day-to-day knowledge of the plant, first- and second-line supervisors in the operations and maintenance departments should help evaluate the safety of any proposed changes.

2.12.2. Examples of Lessons to be Learned from the Failure to Manage Change

Case histories of some catastrophes or near misses that resulted from a failure to adequately manage change are described below.

Case 1: This case shows the necessity of managing change procedures during routine maintenance, such as the installation of rupture discs on leaky pressure relief valves. In a facility, fugitive emissions passing through leaky pressure safety relief valves were stopped with the addition of a rupture disc below the relief valve[5]. However, pinhole leaks in the rupture disc could cause the pressure in the closed space between the rupture disc and the relief valve to rise until it was virtually equal on both sides of the rupture disc. Since rupture discs burst at a predetermined differential pressure, this change in the relief system could seriously compromise the pressure relief capability.

Management of change procedures applied to this proposed pressure relief system change would have suggested a number of additional considerations, ranging from interspace pressure alarm monitoring to special material relief valve seats.

Case 2: This case history demonstrates the necessity of using proper management of change procedures for all applicable changes, including temporary modifications. The most widely publicized catastrophic accident resulting from a temporary modification is the bypass pipe installed at the Nypro plant in Flixborough, UK[6]. The 1974 failure of a badly designed temporary pipe system, used to bypass a reactor that had been taken out for repairs, released hot cyclohexane. The resulting explosion destroyed the plant and killed 28 people. A process modification approval form had been completed in a perfunctory manner.

The Flixborough incident is also an example of the degradation of safety due to a progression of ill advised changes. The carbon steel reactor, which was taken out of service for repair, was damaged by stress corrosion cracking induced by pouring nitrate-containing water over the top stirrer gland to condense leaking cyclohexane vapors. This vapor leak mitigation procedure was not sanctioned by engineering analysis.

Management of change procedures applied to the Flixborough temporary piping modification could have revealed the inadequacy of the temporary piping system and identified proper operating procedures during the temporary operations.

Case 3: This case history demonstrates the need for the indoctrination and training of operations and maintenance personnel in how to recognize situations requiring the implementation of management of change procedures. A sour water separations vessel was equipped with an automatic pump to control level[7]. When the pump failed, the operating procedures indicated that manual level control could be instituted by using an adjacent valve that was piped to the oily water sewer. When this valve was found to be inoperative, a management of change authorization should have been obtained before "jerry-rigging" a tube to drain sour water from the bottom of the sight gauge to the nearby open drain. In this particular case, the operator left the area with the sight gauge drain open and H_2S gas eventually spread throughout the process area.

The lessons learned in this case show the need for recognition of changes that need process change authorization and the need for operations and maintenance personnel fully understand the changes made.

2.12.3. What Constitutes a Change?

Changes that affect a chemical process, other than replacement in kind, include changes to: process chemicals and raw materials; process technology, including production rates, experimentation, and new product development; equipment, including materials of construction, specifications, and process systems; instrumentation, including computer program revisions and changes to alarms and interlocks; operating procedures and practices; facilities, including buildings and containers that affect the process; process conditions and operating parameters; and personnel changes.

The timing of changes can vary from a long-term plan to an emergency condition. Some changes can be extremely subtle or hidden, such as an unusual contaminant in the feed chemicals. Other changes are forced by regulatory directives, such as allowable emissions or waste handling options. Among the most insidious and hard-to-recognize changes are those that result from a series of small changes or from a slow drift in conditions.

Process change is formally defined as a temporary or permanent substitution, alteration, replacement (excluding in kind replacements), or modification by adding or deleting process equipment, applicable codes, process control, catalysts or chemicals, feedstocks, mechanical procedures, electrical procedures, and safety procedures from the present configuration of the process equipment, procedures, or operating limits.

The above definition of process change covers all changes to equipment and procedures. Process changes can range from installing new or additional equipment to adjusting instrument setpoints outside of allowable limits. Table 2-6 gives more examples of process changes that require formal authorization, including safety and environmental review.

TABLE 2-6
Process Change Examples

Examples of Process Change Requiring Authorizations	Examples of Process Change Not Requiring Authorizations
1. Changing the metallurgy of a piping system.	1. Repairs to equipment or piping per established procedures.
2. A piping system change, including adding block valves, new flow paths, etc.	2. Replacement (in kind) of equipment or piping.
3. Installing underground piping, including atmospheric drains.	3. Replacement of equipment or piping components meeting the same specification as original.
4. Installing new pumps or compressors.	4. Cleaning of exchangers, piping or other equipment.
5. Installing structural members to support heavy loads.	5. Area paving.
6. Temporary piping systems.	6. Replacement (in kind) of insulation.
7. Adding or removing insulation to or from process equipment.	7. Painting.
8. Changing the control range of temperature and pressure instruments to exceed defined operating limits.	8. Repairing or recalibrating instruments.
9. Changing or bypassing alarms or permissive switches within an interlock system.	
10. Changing PSV settings or configuration.	
11. Changing the method or control scheme of an instrument loop.	
12. Change in raw materials.	

This is **not** meant to be an "all inclusive" list of examples. If any question arises about whether an authorization is required, begin the authorization procedure.

Critical process equipment, procedures, and operating parameter limits are defined in the operations procedure manuals and the process flow diagram (PFD) or piping and instrumentation diagram (P&ID). In general, in the event of a catastrophic containment failure, critical equipment has the potential to release a significant quantity of flammable or toxic material. Hence, any process equipment or major component that, if it ruptures or otherwise fails, can lead directly to a significant flammable or toxic chemical release is within the scope of this definition. Process piping, pressure vessels, control instrumentation, fire fighting systems, and storage tanks are typical examples of critical process equipment. Water lines, fencing, cooling towers, and water piping corrosion control systems are examples of process facilities which are usually not critical.

Similarly, deviations from predefined process equipment operating limits or changes of written operating procedures constitute a process change that must be approved using the process change authorization procedure. Figure 2-1 contains a detailed example of the potentially hidden safety concerns in a process change.

2.12.4. Process Change Authorization

Process change authorization (PCA) is a formal authorization procedure used to guarantee the effective safety review of appropriate process change requests that will affect process equipment, operating limits, or written procedures.

Process change may be initiated by:

- Work Orders signifying a process change is required.
- Authorization for expenditures (AFEs).
- Process, project, maintenance, or automation engineering requests.
- Suggestions by operating personnel.

Figure 2-2 is a flow chart identifying the major steps needed to execute the process change authorization. As noted in this flowchart, a request for any process change should first be analyzed to determine the need for formal PCA procedures. If process equipment, as previously defined, is involved, the PCA procedures should be used.

The PCA form applicable to each facility provides a checklist-procedure format as well as the signature authority. The level of safety evaluation depends on the details of the process change request. The PCA signatory authority is responsible for ensuring that safety issues have been adequately addressed. If additional technical review is suggested, the authorizing personnel should seek the additional assistance, as appropriate, for the technical nature of the proposed process change.

The documentation for each PCA, regardless of level, should be maintained. Figure 2-3 shows the cover sheet for an example request form. Note that this form requires a list of documents and procedures that will be affected by the proposed process change. After implementation, the documentation

A suggestion to increase the flow rate of a liquid butane line has been made. Economic and schedule considerations suggest that a pump impeller change may be satisfactory.

The new impeller changes the pump characteristics, as shown in the pump curve below, specifically increasing the dead head (zero flow) pressure. The increased flow rate at point B does not change the original design operating pressure at point A.

A safety assessment of the pressure containment capability of the downstream piping and vessels should be completed before this pump impeller change is approved.

Documents reviewed for this assessment include the P&ID, equipment specification sheets, pump curve, pressure relief valve list, piping specification, etc. Heat and Material balance data should be updated. It is determined that no downstream vessels are affected by this equipment change.

In this particular case, the only documents that must change are the pump specification sheet, process flow diagram with heat & material balance, and the P&ID. In some cases, a simple pump changeout may affect many downstream pieces of equipment. Operating procedures may also change.

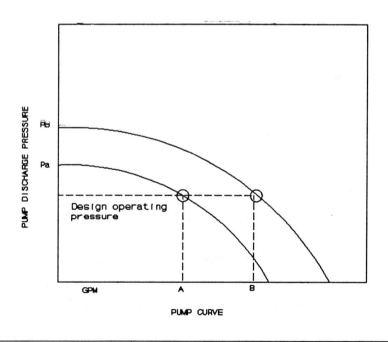

FIGURE 2-1. Process change example.

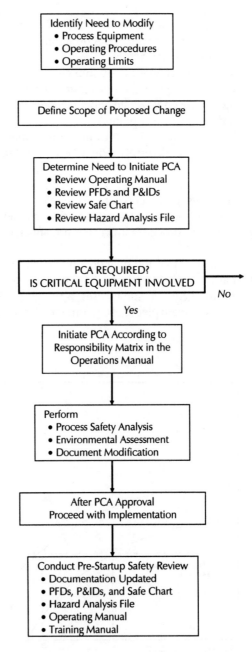

FIGURE 2-2. Example process change authorization procedure.

File No._____   Work Order No._____

  AFE No._____

Description of Change:

Reason for Change:

Process Safety Analysis:

  File No._____ Reviewer:_____ _____ _____
                                          Name   *Title*   *Date*

                      Reviewer:_____ _____ _____
                                          Name   *Title*   *Date*

Elements Considered Concerning Process Change:

  ☐ Environmental   Reviewer:_____ _____ _____
                                          Name   *Title*   *Date*

  ☐ Operations   Reviewer:_____ _____ _____
                                          Name   *Title*   *Date*

  ☐ Operations   Reviewer:_____ _____ _____
                                          Name   *Title*   *Date*

                      Approved:_____ _____ _____
                                          Name   *Title*   *Date*

                      Approved:_____ _____ _____
                                          Name   *Title*   *Date*

List Information Sources Modified (with revision date included:

  Drawings:

  Operations Manual

  Maintenance Manual

  Environmental Permits

  Other (specify)

  Documents attached:

FIGURE 2-3. Example of a major process change authorization form. Temporary or permanent modifications

affected by the process change should be updated during the pre-startup safety review procedure. Information resources that may be changed as the result of a PCA include drawings, operations and/or maintenance manuals, training manuals, and environmental permits.

The personnel authorized to approve PCA forms are usually listed in the facility guidelines or in the standard operating procedures manual. One of the signatory approvals should be from the operations department. The complicated nature of the proposed change or the complexity of the process facility may require higher levels of approval.

The implementation of process change requests should not be permitted until a safety evaluation has been performed. An exception is that emergency work may be completed before formal authorization. An emergency is defined as a situation that poses an immediate threat to health or safety. For process changes not subject to formal PCA procedures, local staff may proceed with the implementation and updating of documents.

2.13 Summary

A key element in a process safety program is a management of change procedure that ensures facilities will remain in compliance with safety objectives, yet recognizes the need for occasional process changes. The purpose of the management of change program is to ensure that safety, loss prevention, environmental, and engineering input are provided at the planning stage of each process change. A formal safety analysis of every critical process change, except replacement in kind or standard maintenance items, is the hallmark of the management of change program. The results of this safety analysis should be disseminated to all affected groups (operations, training, maintenance, process engineering, contractors, etc.) before operations resume. This formal procedure featuring management authorization helps ensure that the individuals initiating and implementing the process change do not overlook a significant point that affects safety or environmental protection. Process safety management regulations[3] and industry recommended practices[4] and guidelines[1,2] suggest that proper management of change will ensure that the process maintains the level of safety defined in the process hazards analysis. The management of change program should be executed primarily by plant operations management staff, process engineering support groups, and first- and second-line supervisors of the operations and maintenance departments. The first- and second-line operations and maintenance supervisors have key roles in executing the management of change program for a facility during the various phases of its life cycle. The discussion in this chapter is intended to provide the reader with the fundamentals of a facility's management of change program. Specific management of change issues that relate to each phase of the life cycle are covered in subsequent chapters.

2.14. REFERENCES

1. Center for Chemical Process Safety (CCPS). *Guidelines for Technical Management of Chemical Process Safety.* American Institute of Chemical Engineers, New York, 1989.
2. Center for Chemical Process Safety (CCPS). *Plant Guidelines for Technical Management of Chemical Process Safety.* American Institute of Chemical Engineers, New York, 1992.
3. *Process Safety Management of Highly Hazardous Chemicals; Explosives and Blasting Agents; Final Rule.* 29 CFR Part 1910, Section 119, Department of Labor, Occupational Safety and Health Administration, Washington, DC, February 24, 1992. *Federal Register,* Vol 57, No. 36, pp. 6356-6417.
4. API Recommended Practice 750, *Management of Process Hazards,* American Petroleum Institute, Washington, D.C., January 1990.
5. T.A. Kletz, *Plant Operations Progress,* vol. 5, no. 3, July 1986, p.136.
6. *The Flixborough Cyclohexane Disaster.* Her Majesty's Stationery Office, London, 1975.
7. *Safety Digest of Lessons Learned.* American Petroleum Institute, Washington DC, 1982.

3

PLANT DESIGN

Process safety should be considered during all phases of plant design. During the design process, the operations and maintenance departments contribute to process safety by informing the designers of the potential hazards that may be encountered during manufacturing. The design and engineering groups need to understand how the plant will be operated and maintained. When the knowledge and experience of the manufacturing personnel are integrated into the design, the resulting plant will not only be safer, but also easier and more efficient to operate and maintain. CCPS has published a number of other guidelines[1-7] which address critical issues with regard to process safety considerations during the design phase.

Plant design can be divided into three phases:[6]

- *Phase I—Conceptual Engineering*—involves the technical and economic evaluation of a project's feasibility, including the process chemistry, process hazards, flow schematics, the fundamental design basis for the equipment, instruments and controls, and safety systems.
- *Phase II—Basic Engineering*—involves process simulation calculations (mass and energy balances) and process flow design, concluding with preliminary piping and instrumentation diagrams (P&IDs), and equipment data sheets issued for design.
- *Phase III—Detail Design*—involves vessel thickness calculations, heat exchanger rating, final P&IDs, line sizing and piping design, and isometric drawings, concluding with specifications and drawings issued for construction.

Not all projects will require all three design phases. For example, a new process will begin with Design Phase I and continue through Phase III. The design for a debottlenecking project would typically start with Phase II and continue through Phase III. A small project such as the relocation of equipment may require only Phase III design. In all cases, the operations and maintenance departments should convey their process safety concerns to the project team as early in the project as possible.

3.1. OPERATIONS AND MAINTENANCE DEPARTMENTS' ROLES

Since the safe operation and maintenance of the plant will become the day-to-day responsibility of the operations and maintenance departments, they assume the role of "client" during the design phase. As the client, the operations and maintenance departments should participate in design reviews and communicate their needs to the project team. To fulfill their responsibility, operations and maintenance representatives should openly contribute their knowledge and experience in process operations and maintenance to the design of the facility. Each piece of equipment, each segment of piping, and all valve configurations should be examined for adequacy and arrangement to safely perform the following functions:

1. Commissioning of equipment
   a. Clean out and wash out
   b. Individual equipment checkout
   c. Leak testing and pressure relief evaluation
   d. Decontamination, water removal, oxygen removal (nitrogen purge)
2. Charging system with inventory
3. Startup
4. Normal operation
5. Shutdown
   a. Normal
   b. Emergency with protective systems activation
   c. With partial loss of utilities or machine failure
6. Decontamination for maintenance
7. Maintenance

As the operations and maintenance personnel review the detail design, they should develop step-by-step procedures for accomplishing each of the above functions. Safety checklists, such as are shown later in this book, or formal HAZOP sessions may be used to examine each of these steps to determine if they can be safely and efficiently carried out. The manufacturing representatives should inform the project team of any changes in equipment and piping or in its placement needed to safely perform these functions. During these reviews, the bases for procedures for commissioning, startup, normal operations, emergency shutdown, normal shutdown and maintenance are developed.

In Phase I, Conceptual Engineering, the operations and maintenance departments should help evaluate process hazards, questioning any inherent process hazards and defining the respective protective systems needed to address these hazards. The operations and maintenance departments play vital roles in achieving an inherently safer design and in controlling hazards to reduce risks. These two issues are discussed in detail in Sections 3.4 and 3.5.

An appropriate time for initiating formal design reviews is near the end of Phase II, basic engineering, when preliminary drawings and specifications are available for overall design review. In Phase III, Detail Design, the manufacturing departments make their greatest contributions by reviewing the design for operability and ease of access with regard to operations and maintenance activities.

An example of an operations and maintenance staff design review is the review of a proposed installation of a multistage cryogenic pump. In considering the startup of that pump, the operations and maintenance personnel would need to determine if the pump could be cooled down without venting to the atmosphere. If venting is required, it would be piped to the flare, as shown in Figure 3-1.

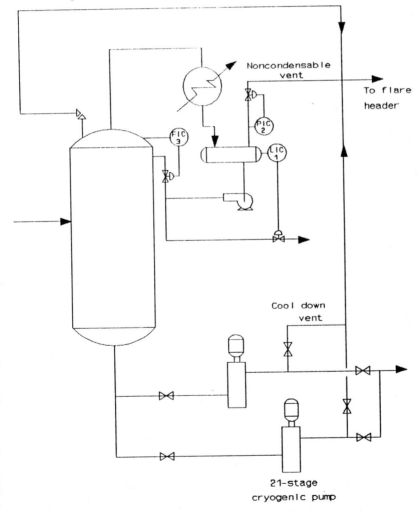

FIGURE 3-1. Importance of design review by operations and maintenance departments.

An example of inadequate operations and maintenance consideration during the design of a vessel is shown in Figure 3-2(a). The flow from the bottom of the vessel is controlled by a pneumatically operated or regulated flow valve, with a bypass valve around the regulated flow valve. Subsequent reviews by operations and maintenance personnel revealed that the level transmitter, control valve, or pump did not have block or drain valves that would allow decontamination or removal for maintenance. Proper operations and maintenance considerations resulted in the design shown in Figure 3-2(b), which includes the necessary valves and piping.

The manufacturing departments should be an integral part of the project team communication network during all phases of the design. For large projects, operations and maintenance representatives may be assigned to the project team. Formal procedures for maintaining communication among the

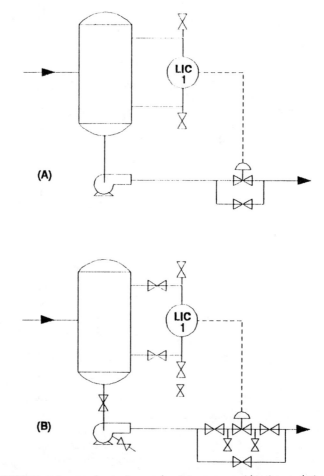

FIGURE 3-2. Impact of operations and maintenance consideration on design.

diverse members of a large project team are vital. For example, in a large international facility, fire hydrants were purchased with bayonet connections while the fire hose was specified with threaded connectors. The resulting inability to respond to the first fire emergency was attributed to a breakdown in communications within the project team. On small projects, formal design communication procedures may not be necessary, as long as the project team maintains open channels of communication.

3.2. DOCUMENTATION

Accurate, complete and up-to-date process documentation is an essential element of process safety management. The project team is responsible for producing, correcting, verifying, and filing documentation of the process and plant design. The project team should ensure that documentation is available when needed. Documentation reviews by the operations and maintenance departments should be conducted in a timely fashion with feedback to the project team.

The operations and maintenance departments need proper documentation and accurate information to safely execute their duties. The importance of preserving and making this information available within a facility includes (a) preserving a record of design conditions and materials of construction for existing equipment, which helps ensure that operations and maintenance staff remain faithful to the original intent; and (b) allowing the rationale for key design decisions made during a major capital project to be recalled, which is useful for a variety of reasons, such as to help design future projects and manage process and equipment changes.

Tables 3-1 and 3-2 show examples of process documentation that should be available to operations and maintenance departments for review during the design and the training of personnel and that should be updated throughout the life of the plant. For example, consider the documentation for a relief valve. Detailed design basis calculations made to determine the pressure relief system specification should be preserved in the event of future system changes.

An example incident in an olefins plant was attributed to a pressure reduction change. The relief valve opened correctly at the new lower pressure, but it did not have sufficient relieving capacity because of the impact of the lower density gases on the flow rate through the flare header piping.

It is important to document communications with the project team, and to track the status and resolution of specific recommendations and issues. The detailed information that contains the basis for developing procedures and manuals for commissioning, startup, operations, shutdown, maintenance, and training should be retained for later use.

TABLE 3-1
process Safety Documentation: Examples of Major
Documents and Files

Operating Procedures Manual

Maintenance Procedures Manual

Project Data Book
    Process Drawing Index
    Equipment Specification Data Sheets

Engineering Standards

Training File
    Manuals
    Records

Emergency Response Plan

Process Safety Management File
    Process Change Authorization Records
    Process Safety Management Procedures Manual
    Safety Audit Reports
    Pre-startup Safety Review Reports
    Hazard Analysis Reports
    Incident Reports
    Mechanical Integrity
        Equipment files
        Inpsection and testing records
        Quality assurance documentation
    Risk Management Plans

Safe Work Practices Manuals

Hazard Communication Handbook

Environmental Permit Files

Documentation requirements for and during plant design, beginning with process definition and design criteria, should be clearly outlined at the very early phases of the project. Copies of documentation should be placed in the plant's permanent process documentation file or process "library." The documentation in the process library should be maintained throughout the plant's life and updated and augmented as necessary. The maintenance of up-to-date documentation is mandated by U.S. federal regulations [29 CFR 1910.119(d)]. Table 3-1 shows the major documents or files containing various portions of the process documentation. These documents are the "master" copies, and operations and maintenance departments should ensure they are continually updated. Table 3-2 shows the required process documentation items that may be contained in one or more of the documents or files listed in Table 3-1.

3.3. PROCESS HAZARD REVIEWS

Process hazard reviews conducted during the design phases are specialized design reviews. Process hazard reviews are organized efforts to identify and analyze the potentially hazardous conditions of a process. Various process hazard review methods are available and each method has its own unique protocol. Specifically, process hazard reviews are used to pinpoint weaknesses in the design and operation of facilities that could lead to accidental chemical releases, fires, or explosions. These reviews provide the project team with information to help them improve the safety of the plant.

TABLE 3-2
Example of Process Safety Information Items

Information Item	Data Source
Process Description	Project Data Book

Process and Mechanical Drawings
Process Flow Diagrams
   Heat and Material Balance Data
Piping and Instrumentation Diagrams
Plot Plans
   Electrical Area Hazard Classification
Electrical One Line Diagrams
	Project Data Book

Safety Protective Systems Relief System Design Basis Process Building Ventilation Data Fire/Gas Monitor Details Passive Protection Systems Safety Interlock Systems	Project Data Book
Process Equipment Specification Sheets Materials of Construction Design Code Compliance—Deviation Report Max/Min Design Limits—Consequence of Deviations Fluid Inventory	Project Data Book
Mechanical Design Data Piping Specifications, general Vessel Specifications, general Electrical Specifications, general Deviations from specifications	Engineering Guidelines and Standards
Operations Procedures Manual Written procedures for each phase Max/Min Operating Limits Consequence of Deviations	Operating Manual

TABLE 3-3
Example of Chemical Hazard Data

General properties	Molecular structure Freezing point Vapor pressure, boiling point Critical pressure, temperature, volume
Physical data	Evaporation rate Vapor density, heat capacity, viscosity, thermal conductivity Liquid density, heat capacity, viscosity, thermal conductivity Latent heats of vaporization and fusion Dielectric constant, electrical conductivity
Fire and explosion hazard data	Flammability classification Flammability limits Explosive limits Flash point Auto-ignition temperature Extinguishing media
Health hazard data	Effects of exposure Permissible exposure limits Immediately dangerous to life and health concentrations
Reactivity data	Chemical stability Decomposition characteristics Polymerization characteristics Effects of temperature and pressure Effects of mixing with other chemicals
Spill or leak procedures	Spill or release response Containment procedures Disposal procedures
Personal protection	Respiratory protection Special clothing Other personal protective equipment Ventilation requirements
Special precautions	Handling procedure Storage requirements Container specifications
Shipping procedures	Container specifications Labeling requirements Shipping procedures

As the departments with ultimate responsibility for process safety management, the operations and maintenance departments should participate in process hazard reviews. Experienced operations and maintenance personnel are essential members of hazard review teams. During process hazard reviews, operations and maintenance personnel should confirm that all required chemical hazards data (see Table 3-3 for example) are available.

The identification of physical changes needed in the plant to improve process safety is one of the goals of process hazard reviews. Figure 3-3 shows the various stages of a project when process hazard reviews may be performed, starting with the earliest conceptual considerations. Ongoing process hazard reviews should be considered an integral part of each phase of the life cycle of a plant; however, physical changes made during the detail design phase are often most effective and less expensive to implement than after construction or fabrication begins. Process hazard review methods depend on the level of information available on the project. In the conceptual engineering phase, only general information is available, such as process chemistry and estimates of product and by-product inventories, flow rates, temperatures, and pressures. During this phase, only preliminary hazard assessments can be performed. As the project reaches the end of the basic engineering phase, a second process hazard review is usually performed. The objective of this review is to ensure that the major process hazards have been identified and that necessary corrective recommendations are implemented. A third process hazard review should be completed before construction contracts are awarded. This process hazard review is a design verification review and is held just before all design bid packages are completed. Its purpose is to ensure

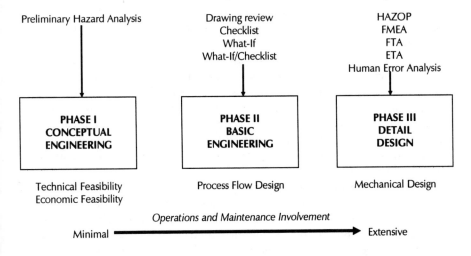

FIGURE 3-3. Timing for hazard evaluations during major process unit design.

that all items relating to previous process hazard reviews have been resolved
and that recommended equipment changes are included in the bid package.
Examples of issues[6] typically covered in process hazard reviews are:

- Equipment design temperature and pressure vs. process conditions
- Pressure-relief device sizing criteria and locations
- Control valve "fail-safe" mode
- Set points of emergency alarms, trips, and interlocks
- Reactive chemical considerations
- Backup provisions for utilities and services
- Computer control programs
- Electrical classifications
- Critical instruments
- Fire protection and vapor suppression system design
- Vent header and flare system sizing criteria and compatibility of pres-
  sure-relief device tie-ins
- Compatibility of materials of construction
- Line sizing and specification
- Equipment arrangement

Many publications[7,8] describe methods and examples for conducting of
the various process hazard review procedures, the most common of which are
summarized in Table 3-4. As shown in the table, the degree of operations and
maintenance participation in the different process hazard reviews depends on
the particular review. For example, HAZOP is one of the process hazard
review methods in which operations and maintenance representatives have
an extensive and determining role. The HAZOP technique is a multidisciplinary approach based on the principle that several experts with different
backgrounds can identify more problems when working together than when
they work separately and combine their results. During a series of meetings,
the multidisciplinary HAZOP team (which includes operations and maintenance personnel) reviews process drawings or procedures using guide words
to find possible deviations from the design intent. The first step in a HAZOP
study is the selection of study areas called nodes. Selected guide words (e.g.,
more, less, etc.) are applied to the process parameters (e.g., flow, pressure, etc.)
at each node to find possible deviations from the design intent. The causes and
consequences of deviations are determined, and, finally, safeguards or recommendations are made for design provisions, procedural controls, or areas that
need further study.

In contrast, fault-tree analysis is a process hazard review method in which
the operations and maintenance representatives have a comparatively minimal role. For example, consider the opening of a pressure relief valve and the
release of hazardous fluid to the environment. The release of the hazardous
fluid through the relief valve is the top event. The fault tree construction begins
with the top event and proceeds level by level until all faults have been traced

TABLE 3-4
Different Process Hazard Review Methods and Degree of
Operations and Maintenance Participation

Method	Design Phase	Data Requirements	Results	Example	Operations/ Maintenance Participation
Preliminary Hazards Analysis	I	Chemicals and their properties, equipment, operationg conditions	Qualitative description and ranking of hazards	Feasibility study of a new chemical plant	Minimal
Drawing Review	II/III	PFD, P&ID, Isometrics and pipings	List of comments, recommendations, and questions regarding the safety and operability of the design	Review of vent and flare system for light gas plant	Extensive
Checklist Analysis	II/III	Equipment specifications, operating procedures, system documentation	Compliance with prescribed procedures, verification of operating conditions, or component operating ststus	Review of refrigeration compressor interlock system on ethane/ ethylene splitter	Moderate to Extensive
What-If Analysis	II/III	Chemicals and their properties, design intent, process description, material handling, operation and maintenance precedures	Tabular compilation of what-if scenarios with consequences, safeguards, and recommendations	Overfilling of cryogenic tank	Moderate to Extensive
What-If/ Checklist Analysis	II/III	Combination of what-if and checklist analysis as listed above	Listing of potential accident situations, effects, safeguards, and recommendations	Impact on flare system from operation of cryogenic relief valve on ethane/ ethylne splitter	Moderate to Extensive
HAZOP	III	PFDs, P&IDs, logic diagrams, process description, operating procedures	Report containing HAZOP worksheets and summary of recommendations	Pre-startup HAZOP of booster compressors for sour gas service	Extensive
Failure Modes and Effects Analysis	III	Equipment description, design intent, operating procedures	Tabular compilation of equipment with corresponding failure modes and effects	FMEA study to address safety hazards of DAP reaction system	Moderate to Extensive

Continued on page 44

TABLE 3-4
Different Process Hazard Review Methods and Degree of
Operations and Maintenance Participation

Method	Design Phase	Data Requirements	Results	Example	Operations/ Maintenance Participation
Fault Tree Analysis	III	Failure scenario, equipment and process description, operating and emergency procedures	Report containing fault tree for the specified scenario and recommendations	FTA of events leading to the release of hazardous fluid srom a relief valve	Minimal
Event Tree Analysis	III	Initiating event, equipment and process description, operating procedures	Report containing event tree for the specified scenario and recommendations	ETA	Minimal
Cause– Conse- quence Analysis	III	Accident scenario, equipment and process description, operating and emergency procedures	Report containing cause–consequence diagrams, discussion of the consequences, and recommendations	Cause–consequence analysis of loss of cooling water to reactor	Minimal
Human Reliability Analysis	III	Process and equipment description, operating procedures, layout, training programs	Report containing a list of corrective actions that will reduce the likelihood of human errors	Human reliability analysis of operator response to an alarm	Minimal

to their basic causes. Operations and maintenance representatives may identify top events for a fault-tree analysis during process hazard reviews; however, because of the special skills needed, in most cases, the actual fault-tree
analysis will be conducted by other professionals.

Design changes made after a hazard review should not escape re-review.
Management of change techniques should be employed for design changes
made during engineering and construction.

3.4. DESIGNING FOR INHERENT PROCESS SAFETY

Inherently safer chemical process designs concentrate on avoiding process
hazards rather than on controlling them by adding protective equipment and
procedures. Opportunities for achieving inherent process safety are primarily
available during Phase I, Conceptual Engineering. However, process modification projects may also provide an opportunity to use inherent process safety

concepts. Selecting safe process fluids, minimizing inventories, and choosing the safest possible operating and storage conditions are some of the more important ways that process safety features can be incorporated into the plant design.

3.4.1 Process Fluids

By applying the idea of inherently safer design, the hazards of raw materials or intermediate (in-process) products can be reduced or eliminated. The elimination or reduction of hazardous intermediates may be done during the design phase. The role of the manufacturing departments should be to evaluate any trade-offs.

Sometimes, a slightly different process may require less hazardous intermediates. The intermediate product, methyl isocyanate[11,12], which leaked and killed more than 2,000 people in Bhopal was used to manufacture the insecticide carbaryl. In an alternate process, the same final product can be manufactured using the much less hazardous intermediate product chloroformate ester instead of methyl isocyanate[12].

It is also sometimes possible to replace a process fluid with a less hazardous material. An example is using water instead of kerosene for cooling the catalyst tubes in ethylene oxide plants. Another example is using glycol-water mixtures in cryogenic vaporizers, replacing the flammable isopentane heat transfer medium.

3.4.2. Inventory Minimization

The inventories of flammable or toxic fluids should be minimized. With the reduction of hazardous materials, the size of the various units (i.e., reactors, distillation columns, heat exchangers, etc.) also can be reduced. This size reduction, in turn, can lead to simplifying the controls. For example, a larger vessel may require a staged pressure relief valve arrangement to safely relieve the pressure. A smaller vessel would require only a single relief valve to protect the vessel's integrity.

A costly lesson of the Bhopal accident[11–13] has been to design processes in which large inventories of hazardous intermediates are not retained. The design should explore all possible ways of reducing the quantities of hazardous intermediates by reacting them as they are produced. This may mean using a continuous flow operation instead of batch flow. Thus, the manufacturing departments can help set "minimum operability" limits.

3.4.3. Operating and Storage Conditions

Extensive consideration should be given to options that allow a less hazardous operation of the plant. Some generic examples are: carrying out a hazardous process under less hazardous conditions (e.g., lower pressure or lower temperature), or storing or transporting a hazardous material in conditions close to atmospheric. Flashing fluids are under pressure above their atmospheric pressure boiling point, and on loss of containment they produce large quantities of vapor and aerosol. In contrast, liquids below their boiling points produce very little vapor. The objective is to reduce the severity of the operating conditions, which will, in turn, reduce the severity of any potential incidents.

In some instances, replacing a hazardous fluid or changing operating conditions will require additional process equipment. The additional complexity created by using the replacement fluids should be compared with the complexity of original design. The nitration of aromatic hydrocarbons, for example, is often accomplished in batch reactors at high temperatures. The operation is hazardous because of the potential for runaway reactions at those temperatures[12,13]. The operation is much safer if the reaction mixture is diluted by a solvent; however, additional or larger reactors are needed to obtain the same yield.

3.5. CONTROLLING OF HAZARDS TO REDUCE RISKS

The project team and the operations and maintenance departments must be satisfied that hazards that cannot be eliminated are controlled. An acceptable level of risk is achieved by a design that reduces the likelihood of a hazardous event occurring or that mitigates the consequences of such an event. Following the process hazards analysis in which operations and maintenance personnel participate, the project team should do a risk assessment. Sometimes, the risk assessment may be a qualitative judgement based on project team experience. Depending on the potential severity of the consequences, more quantitative risk assessment techniques may be needed.

If the risk is judged to be unacceptable, risk reduction techniques should be employed. Risk reduction measures usually are incorporated at the end of the basic engineering phase and throughout the detail design phase, when the operations and maintenance departments have the most interaction with the project team. The manufacturing departments must be satisfied that the hazards are reduced to acceptable levels for safe operation of the plant. As part of the process hazard review, operations and maintenance staff may recommend safety protective systems such as barricades, gas detectors, emergency isolation valves, trips and alarms, deluge systems, and operating restraints that can reduce risks.

Several types of multiple safeguard techniques may be considered to reduce risks. Multiple safeguard techniques are generally classified as either containment integrity or consequence minimization techniques. The manufacturing departments can and should play a key role here by helping to evaluate

TABLE 3-5 Examples of Containment Integrity Techniques	
Process design	Codes, standards, recommended practices Manufacturing and construction quality assurance Process unit redundancies
Operational controls	Process control within predetermined limits Safe work practices Operator training Operating procedures manual
Safety protective systems	Emergency shutdown Pressure relief devices—to vent or flare Isolation valves Uninterruptible power supply—backup power for control system Vehicle barricades

the effectiveness of active versus passive protection and redundant protection necessary in the system.

Table 3-5 gives examples of containment integrity techniques. For each technique, some special process equipment, safety protective systems, preventive maintenance practices, or restrictive operating procedures may be needed.

Table 3-6 provides examples of consequence minimization techniques, which are generally considered the final protective barrier. Therefore, containment integrity techniques are the primary process risk reduction technique that should be used, and the consequence minimization techniques should be added as final backup systems.

3.6. PLANT LAYOUT

Planning the plant layout consists of two parts: planning the overall site and planning the process area layout. Site planning consists of locating the storage facilities, process areas, utilities, laboratories, flares, docks, loading areas, control rooms, administrative offices, roads, etc. Process area layout concerns the arrangement and placing of equipment and piping within a process area. The operations and maintenance departments may be asked for input about site planning; however, their major contribution to the design comes in planning the process area layout, which has a major effect on the safe day-to-day operation and maintenance of the process units. The operations and maintenance departments should communicate to the project team "how they will operate the plant, i.e., to station operators, control rooms, etc."

TABLE 3-6	
Examples of Consequence Minimization Techniques	
Incident detection	Ambient air monitors with alarms Flammable gas monitors with alarms Heat and flame detectors with alarms
Early response	Automatic operation of sprinkler, foam, deluge, steam snuffing system
Passive protection	Separation distance between equipment and operating units Diking/grading Special fire protection (insulation/ablative) Blast resistant control rooms Buried pipelines Ignition source control (fixed and mobile)

Emergency response	Fire responder training for operations and maintenance personnel on
	    fixed and mobile fire-fighting and response equipment (e.g., fire
	    water, foam, dry chemical)
	Emergency response plans and training
	Regular practice drills

Fire containment and response team	Designated in-plant fire brigade Routine fire suppression training Agreements with local fire departments to respond Industrial mutual aid organization response Mutual assistance organizations

3.6.1. Site Planning

Site planning should cover the process safety of the plant, as well as the safety of surrounding units, control rooms, and the communities outside the plant boundaries. Table 3-7 lists some process safety considerations useful for site planning. The design basis and assumptions for the site plan and plant layout documents should be compiled into a site plan document or integrated with another relevant plant document. Changes or additions made to the plant afterward should conform to the site plan. Deviations from the overall site plan should be made only after a detailed hazards analysis and consideration of the effect on the rest of the plant.

The plant should have as few obstructions to vehicles as possible[13,14]. Turning areas and road bed integrity should be considered for the movement, placement, and removal of heavy equipment by cranes. Roads should be designed to allow two vehicles to pass easily. If possible, each area should be accessible from at least two different routes. Ease of access and egress is not only useful during emergencies but also makes the job of operating and maintaining the plant much easier and safer. In every design, and especially irregular terrain, the effect of out-of-control vehicles should be taken into

TABLE 3-7
Site Planning Safety Considerations

Issue	Description	Operations and Maintenance Involvement
Separation Distances	Process units, storage tanks, buildings, utilities, etc. should be separated enough that the probability of the domino effect is reduced.	Participatory
Critical Equipment	The locations of and distances between critical equipment should receive special attention.	Participatory
High-Hazard Areas	Similar to critical equipment, the locations and distances between high-hazard areas should receive special attention.	Participatory
Utilities	The layout of utilities, such as water, steam, and electricity should be planned so that they can be maintained as well as possible during potential incidents and emergencies.	Participatory
Control Room(s)	Control rooms(s) should be located so that these areas are the last to be compromised in case of an incident.	Lead
Instrumentation and Controls	The location of instrumentation and controls should allow the best possible access during emergencies.	Participatory
Flooding and Drainage	Protection against flooding of critical areas and adequate drainage throughout the plant.	None
Local Conditions	Consideration of conditions specific to the area, e.g., severe storms, hurricanes or earthquakes.	None
Adjacent Land	Effect of activities on the adjacent land on the activities in the plant and vice versa.	None
Emergency Activities	Ease of access for emergency crews. Emergency equipment, fire-fighting, communications.	Supportive
Escape Routes	Easy escape routes for all workers during emergencies.	Participatory
Plant Boundaries	Demarcation and ease of control.	None

consideration during site planning. The operations and maintenance departments can help select travel routes.

The hazard of a process unit being ignited by a fire in an adjoining area is the main consideration when determining separation distances[15]. The objective of proper spacing is to contain the fire in the area where it originates. In addition, the size of the areas should allow for effective fire fighting. The operations and maintenance departments should help evaluate the size of the areas. Flares can become a source of igniting vapor released from other areas of the plant. In addition, flares can go out and create a hazardous condition by discharging flammable vapors; therefore, it is important that flares be located in a remote area of the plant and that the prevailing wind direction and topography be given appropriate consideration.

It may be desirable for utilities to remain accessible and impregnable even under severe emergency conditions. Thus, considerable thought should be given to reducing the exposure and subdivision of utilities so that independent services of steam, electricity, water, etc. can be maintained. Electric power should be provided to a given area from two directions. Plants relying on outside power should have two separate feeder circuits entering the plant. The locations of fire water pumping stations are determined on the site plan. The fire water system should be looped so that a given hydrant can be fed from two different directions. These considerations also apply to utility steam as well as to water.

3.6.2. Process Area Layout

Process area layout is one of the main design considerations that can affect the safety of each of the functions that operations and maintenance must perform during the life of the plant. The layout of a process area is determined primarily by the requirements for its efficient functioning during daily activities. The manufacturing departments should explain how the process area can best be arranged to accommodate commissioning, startup, normal operations, emergency shutdown, normal shutdown, and maintenance. For example, the physical arrangement of equipment directly affects safety. The arrangement of shell and tube heat exchangers and the associated lay-down area necessary for cleaning operations is a key issue for most operations and maintenance departments. Consider the effect of process area layout in a recent situation involving excessive damage to an organic intermediates plant. In a plant with insufficient spacing, the emergency response route for fire equipment was blocked by a tube bundle that had been taken out of service for cleaning.

The layout of process areas is predominantly controlled by such considerations as engineering standards, insurance criteria, and maintenance engineering considerations. Operations and maintenance personnel can play an important role in plant layout as it pertains to accessibility, operability, and maintainability, where the distinction between process safety, personnel safety, operability and maintainability is not exact because a single issue can affect several areas.

Table 3-8 lists a series of questions that operations and maintenance personnel can use as a guide when reviewing proposed process area configurations for adequate access as it affects process safety, personnel safety, operability, and maintainability.

TABLE 3-8
Examples of Plant Layout Considerations with Regard to Process Safety,
Personnel Safety, Operability and Maintainability

1. Are utility stations (steam, water, and air) arranged to keep aisles and operating areas clear of hoses? Are there adequate numbers to provide access to all operating areas?

2. Are operations in enclosed buildings or structures (i.e., analyzer houses) ventilated to prevent buildup of vapors, dusts, and excessive heat?

3. Where compressed gas bottles are required for analyzers or other operations, are restraining stands provided?

4. Are controls housed in separate structures for potentially hazardous operations? Can they be locked out for equipment maintenance?

5. Do platforms provide adequate access for safe operation and maintenance of elevated equipment?

6. Are nozzles and manways on tanks, columns, and other vessels sized and located for safe entry and exit during cleaning and maintenance?

7. Is protection provided against contact with hot surfaces?

8. Is head clearance adequate in walkway and working areas?

9. Is power-driven equipment adequately spaced and rotating elements guarded?

10. Are manually operated valves, switches, and other controls readily accessible to the operator from a safe location? Can they be locked out according to plant lock/tag procedures for maintenance of the equipment? Are elevated valves provided with platforms, chain, or remote operators?

11. Is equipment (lines, columns, exchangers, pumps, compressors, etc.) provided with high and low point vents and drains so they can be drained and decontaminated for maintenance?

12. Are vents and drains located where they are accessible for operation? Where the equipment contains flammable, corrosive or toxic materials, are the vents and drains connected to a vent, flare or drain system to prevent release to the environment during draining and decontamination?

13. Are emergency showers and eye baths provided? Are they accessible? Are there adequate numbers for all operating areas? Are they provided with alarms? In remote operating areas, do the alarms sound in a control room or some other constantly attended location?

14. Has a safe storage and dispensing location for flammable liquid drums been provided? Is grounding and bonding provided? Is the distance from ignition sources adequate?

15. Can dual train equipment (pumps, compressors, exchangers, etc.) be adequately isolated, drained, and decontaminated for maintenance during operations? In processes containing hot, flammable, corrosive or toxic materials, are double block and bleed valves provided? If they are in fouling service, is the design of the valves such that they will be operable and provide tight shutoff?

16. For toxic and flammable materials, is the area curbed and drained so that a spill will not pool under the equipment, hindering emergency response?

Continued on page 52

TABLE 3-8 *(Continued)*
Examples of Plant Layout Considerations with Regard to Process Safety,
Personnel Safety, Operability and Maintainability

17. Do critical control valves have manual by-passes to allow for control of the process if the valve fails or must be removed for maintenance?
18. Is there adequate overhead clearance over pumps and compressors for a lift or hoist if they have to be removed for maintenance?
19. Is there adequate clearance for pulling bundles from exchangers for maintenance or cleaning? Have dropped object problems been given due consideration?
20. Will protective systems be separate from basic control systems for additional security from common-cause failures?
21. If foul weather shelters are provided for field operators, are they located so that the operating areas are visible?

3.7. PLANT STANDARDS AND PRACTICES

The various equipment and specifications for a processing plant today may include references to standards produced by more than 20 separate organizations.

Table 3-9 lists various organizations that have published standards and recommended practices for various processing plant equipment or operations and maintenance practices. In addition to process safety regulations and industry-published standards and recommended practices, plant design is subject to internal company guidelines. The operations and maintenance departments will have knowledge of undocumented, but beneficial, plant practices that can influence plant design; these should be communicated to design personnel. Operations and maintenance personnel can help prepare and revise plant standards and practices to improve process safety.

Continuous vigilance is an integral part of the activity of the manufacturing departments because it is responsible for ensuring adherence to corporate operations and maintenance practices. Deviation from accepted corporate operations and maintenance guidelines can result in catastrophic loss. The inadequate lock out/tag out procedure used by the operations and maintenance departments on the Piper Alpha production platform ultimately led to destruction of the platform and 167 lives lost. An electrical lockout of a standby propane pump was not properly communicated to the operations department. Following a shift change and a failure of the primary propane pump, the attempt to start the standby pump (which was not adequately isolated) led to a destructive propane release.

TABLE 3-9
Sources of Process Plant Standards and Practices

1. Testing Standards and Safety Groups
American National Standards Institute (ANSI)
American Society for Testing and Materials (ASTM)
National Fire Protection Association (NFPA)
Underwriters Laboratories, Inc. (UL)
National Safety Council (NSC)
National Board of Boiler and Pressure Vessel Inspectors (NBBI)

2. Insuring Associations
American Insurance Association (AIA)
Factory Insurance Association (FIA)
Factory Mutual System (FM)
Industrial Risk Insurers (IRI)
Oil Insurance Association (OIA)

3. Professional Societies
American Conference of Governmental and Industrial Hygienists (ACGIH)
American Industrial Hygiene Association (AIHA)
American Institute of Chemical Engineers (AIChE)
American Society of Mechanical Engineers (ASME)
American Society of Heating, Refrigeration, and Air-Conditioning Engineers (ASHRAE)
American Society of Safety Engineers (ASSE)
Illumination Engineers Society (IES)
Institution of Electrical and Electronics Engineers (IEEE)
Instrument Society of America (ISA)
Systems Safety Society (SSS)

4. Technical and Trade Groups
American Waterworks Association (AWWA)
Air Conditioning and Refrigeration Institute (ARI)
American Gas Association (AGA)
American Petroleum Institute (API)
Chlorine Institute (CI)
Compressed Gas Association (CGA)
Tubular Exchangers Manufacturers Association (TEMA)

5. United States Government Agencies
Bureau of Mines (BM)
Department of Transportation (DOT)
U.S. Coast Guard (USCG)
Hazardous Materials Regulation Board (HMRB)
Environmental Protection Agency (EPA)
National Institute of Standards and Technology (NIST)
Occupational Safety and Health Administration (OSHA)

3.8. HUMAN FACTORS IN PLANT DESIGN

The human factors that affect plant design concern the skills and training of the personnel, the operating and maintenance procedures, the design of the equipment, and the environment. The operations and maintenance departments are responsible for ensuring that new facilities are designed so they can be safely and efficiently operated and maintained by plant personnel. To operate and maintain a plant safely, human factors should be considered in every phase of the design. Ergonomics is a word that has become popular in recent years. A system or item of equipment (or plant design) that is ergonomically sound is one that allows humans to perform their tasks safely and efficiently. The interface of human beings and equipment should be considered and the concept of ergonomics applied at every point in the design. Often compromises must be made. For example, a change in design to improve the operation could make maintenance more difficult or hazardous.

Human reliability is another factor that affects safety and should be considered during the design process. Humans can make errors when performing tasks and making judgements, particularly when performing complex operations under stressful conditions. Emergency shutdowns caused by equipment failure can be very stressful, and if the procedures, equipment, and control layout are confusing, errors can easily be made. An example (discussed earlier in section 2.6) of this occurred when a cracking furnace in an olefins plant ruptured a tube and caught on fire. The board operator shut down the wrong cracking furnace because of the confusing layout of emergency shutdown switches. During the design phase, the experience of operations and maintenance personnel can help reduce or eliminate opportunities for this type of error.

Human reliability analysis is a systematic evaluation of the factors that influence the behavior and performance of plant personnel. The analysis looks at the physical and environmental factors involved in a job and the skills, training level, knowledge, etc. of the personnel. This analysis is used to locate areas or situations in which the person in charge could make an improper decision that leads to an incident. The information needed for a human reliability analysis includes normal and emergency operating procedures, a knowledge of employee training levels, the layout of control and alarm panels, and job descriptions for the personnel.

For example, consider an analysis of operator response to an alarm in the control room. The human reliability analysis technique might look into factors such as excessive noise in the control room, the location of the shutdown switch, visual indicators for easy identification of the beginning of a shutdown, and the training and background of the operators. The result of the analysis is a list of human errors that could lead to a hazardous situation and recommended corrective actions for each error.

Many catastrophes or near misses have occurred because of operating and maintenance personnel error[9,16]. These errors could have been avoided if potential problems had been investigated during the design phase and possible ways for preventing or overcoming them had been built into the design. The checklist of potential human reliability problem areas given in Table 3-10 should be considered by operations and maintenance personnel when they participate in design reviews.

Several types of incidents caused by human error can be avoided by eliminating design weaknesses. Take for example a mechanic walking from his work bench and falling into an open vehicle inspection pit, or a technician interchanging the inlet and outlet valves of a compressor. Both incidents can be avoided by paying proper attention during the design process. For example, the open pit problem can be avoided by proper layout and by placing barriers that make it impossible to stumble into the pit. The compressor problem could be solved easily by making inlet and outlet valves of different size or appearance.

3.9. MAINTENANCE CONSIDERATIONS

The maintenance requirements of a process plant are set by initial installation choices made during the design phase. The designers of process plants can eliminate or reduce many potential maintenance hazards by engineering them out of their designs. Besides cutting maintenance costs, proper design can make it safer and easier to maintain process plants. The operating and maintenance departments can ensure that new facilities are designed to allow safe, effective, and efficient maintenance by informing the project team of maintenance requirements and by working with them to ensure that their ideas are addressed. Designers can then enhance safety by providing ample access when laying out an operating area, by requiring that manufacturers include maintenance aids in their equipment designs, by providing the working space needed for maintenance, and by specifying built-in handling facilities where they may be needed. Checklists can be used during design reviews to ensure that maintenance needs are not overlooked. The project team should seek the answers to such questions from the operations and maintenance departments. Other maintenance considerations that may affect process safety include reliability and availability, redundancy, diversity of redundant systems, and proof-testing procedures and provisions.

For easy maintenance in a process unit, similar items of equipment should be near each other when such an arrangement would not adversely affect the safety of operations. For example, pumps can be grouped so they can be easily reached and handled. Whenever possible, pumps should be located outdoors and not inside buildings. Outdoors, pumps should be located next to an access way and out from under low-clearance structures. Pumps are much easier and

TABLE 3-10 Example Human Engineering Checklist	
Human–machine interface	Adequate labeling Proper arrangement/placement Adequate displays Adequate controls Ease of monitoring Unit differences Color selection
Work environment	Housekeeping Hot/cold environment Lighting Noise Cramped quarters
Complex systems	Operating procedures and training Monitoring multiple items
Fault-tolerant system	Errors not detectable Errors not recoverable

safer to work on when clear space is provided around each pump. The maintenance department should help the project team examine proposed equipment locations.

Practical maintenance department input at the design stage to promote inherent and ongoing mechanical integrity can provide and unique and major contribution to process safety. Plant equipment may be laid out so that it meets all the safety criteria for normal operations, but it still may be hazardous during maintenance. For example, heat exchangers should be oriented so that during cleaning and the removal of tube bundles, the associated lifting equipment will not impede emergency access.

3.10. MANAGEMENT OF CHANGE

After operations and maintenance have communicated their requirements to the project team, they should follow up during the design process. Operations and maintenance personnel should carefully review engineering drawings, bid packages, and other engineering documents to ensure that their requirements have been incorporated into the facility design and specifications. The emphasis of these reviews is process safety. The operating and maintenance department should ensure that all appropriate process hazard reviews are being conducted at the various stages of the project.

The administrative procedure used by many design projects to modify the design detail or add a newly recognized feature is the "change order." The design project change order should incorporate the management of change concepts described in Chapter 2. Small changes initiated by internal engineering design staff, which may not follow the change order procedures, should follow the management of change authorization procedures. Once a process hazard review has been formally issued, any design changes, alterations, modifications, deletions, etc., should be reviewed for their effect on safety using management of change procedures.

The recent experience of an LNG facility emphasizes the importance of managing change during design. The project team specified a fire protection system that would fill the earthen dike around a large grade-level cryogenic tank with high expansion foam. After the process hazard review and final specification of the foam system, a piping pressure drop analysis resulted in a design change that elevated the cryogenic tank on pylons 10 feet above ground. The effect of this change was not discovered until the fire protection system startup team arrived on site long after the tanks were constructed. A major redesign of the foam system had to be undertaken because of the change from grade-level to elevated configuration of the cryogenic tank it was now expected to protect. A management of change procedure would have alerted the operations and maintenance departments to the tank design modification, resulting in a cost-effective construction project.

3.11. REFERENCES

1. Center for Chemical Process Safety. *Guidelines for Engineering Design for Process Safety.* New York: AIChE, 1993.
2. Center for Chemical Process Safety. *Guidelines for Preventing Human Error in Process Safety.* New York: AIChE, 1993.
3. Center for Chemical Process Safety. *Guidelines for Safe Automation of Chemical Processes.* New York: AIChE, 1992.
4. Center for Chemical Process Safety. *Guidelines for Vapor Release Mitigation.* New York: AIChE, 1988.
5. Center for Chemical Process Safety. *Guidelines for Safe Storage and Handling of High Toxic Hazard Materials.* New York: AIChE, 1988.
6. Center for Chemical Process Safety. *Guidelines for Technical Management of Chemical Process Safety.* New York: AIChE, 1989.
7. Center for Chemical Process Safety. *Guidelines for Hazard Evaluation Procedures.* 2nd ed., New York: AIChE, 1992.
8. Greenberg, H.R., and Cramer, J.J., eds. *Risk Assessment and Risk Management for the Chemical Process Industry.* van Nostrand Reinhold, New York, 1991.
9. Kletz, T.A. *Learning from Accidents in History.* Butterworths, Tonbridge, UK, 1988.
10. Kletz, T.A. *Plant Design for Safety: A User Friendly Approach.* Hemisphere Publishing Corporation, New York, New York, 1991.

11. *Bhopal Methyl Isocyanate Incident: Investigation Team Report.* Union Carbide Corporation, March 1985.
12. Bretherick, L. *A Handbook of Reactive Chemical Hazards.* 3d ed. Butterworths, Tonbridge, UK, 1985.
13. Burklin, C.R. "Safety Standards, Codes and Practices for Plant Design." *Chemical Engineering, Albany,* vol. 79, p. 56, Oct. 2, 1972.
14. *International Symposium on Preventing Major Chemical Accidents.* American Institute of Chemical Engineers, New York, 1987.
15. *Recommended Spacing for Safety in Chemical Facilities.* Industrial Risk Insurers, New York, 1976.
16. Center for Chemical Process Safety. *Guidelines for Investigating Chemical Process Accidents.* New York: AIChE, 1992.

4

PLANT CONSTRUCTION

Achieving process safety by controlling hazards and reducing the consequences of an incident can be accomplished by maintaining process and equipment integrity. Decisions about the maintenance of process and equipment integrity are usually made during the design phase. However, during the construction phase, the design factors that affect process and equipment integrity can be verified and expanded. The operations and maintenance departments can play a vital role in ensuring that equipment integrity is properly considered during construction. They can help ensure that the construction is carried out according to specifications and that the plant will safely contain and control the materials being processed under all conditions of service. Also, construction is the time when operations and maintenance departments gather information for operating procedures and data for mechanical integrity programs.

The role of the operations and maintenance departments can be extensive during major construction changes to an existing process plant. On the other hand, during turnkey new plant construction, their roles may be minimal. The material presented in this chapter applies mainly to the roles played by the operations and maintenance departments during major modifications (e.g., the addition of a new unit) to an existing plant. The degree of involvement of the operations and maintenance departments, as defined earlier in Chapter 2, increases gradually during the design phase until construction starts. During the construction phase, the degree of involvement remains at a constant level and consists of oversight, assistance with inspections, and data gathering for operating procedures and mechanical integrity programs.

4.1. ROLES OF THE OPERATIONS AND MAINTENANCE DEPARTMENT

The roles of the operations and maintenance department during the design phase continue to increase gradually. When construction starts, the operations and maintenance degree of involvement increases steeply. The oversight role continues during construction, where the process safety management focus becomes quality assurance—to ensure that the process equipment is built according to design specifications, that it is assembled and installed properly, and that the maintenance department has the required information, training, and tools to maintain the original level of equipment integrity. Although the immediate responsibility for quality assurance belongs to the project manager, operations and maintenance personnel can provide valuable assistance with equipment inspection and quality control. It is not uncommon for operations and maintenance personnel, because of their work background and experience, to identify safety hazards, operability, and maintainability problems that may have passed undetected by the engineering groups in the design, drafting, and specification stages. In addition to oversight during the construction phase, the operations department starts gathering information for developing operating procedures, and the maintenance department starts gathering equipment data for recordkeeping and developing maintenance procedures.

Plant management should make sure that operations and maintenance personnel are involved in the plant construction and quality assurance procedures as much as possible. Thus, operations and maintenance personnel can not only help with the verification and inspection process, but also start learning about activities related to the startup, operation, and maintenance of the plant. Many startup and operating incidents attributed to human error can be avoided by involving operations and maintenance personnel during construction.

4.1.1. Communication and Coordination with Project Team

Communication (both formal and informal) between the project team and the operations and maintenance departments should continue as established during the project design phase. Typically, the project manager is the operations and maintenance departments' interface with construction. Formal communications should include a written schedule of construction activities established before any construction work begins. In addition to specifying dates for beginning and completing construction activities, the schedule should clearly identify the duties and responsibilities of each department. After construction has started, the project team should hold regular meetings with operations and maintenance department representatives to review the status of construction, revise the construction schedule, and discuss upcoming activities. In addition to avoiding incidents, communication among the different departments involved reduces construction, pre-startup, and commissioning delays. Table 4-1 gives examples of construction activities where the coordination of different groups can reduce construction delays and at the same time help avoid hazardous situations.

For example, because of excellent coordination and cooperation between the project team and the operations and maintenance departments, a major expansion of a herbicide plant was completed two months ahead of schedule.

Scheduling meetings were held every week on Mondays and Thursdays. The project team found ways to do things more quickly by coordinating its activities with the operations and maintenance departments. A very important benefit was that the operations and maintenance departments were able to provide their input to the project team and plan pre-startup and commissioning tasks.

In addition to the formal communications described above, it is very important that informal communications between the project team and the operations and maintenance departments be maintained throughout construction. The project manager should ensure that the project team is familiar with the operations and maintenance first- and second-line supervisors. The area supervisor (both operations and maintenance) should regularly be in touch with the construction supervisor to avoid scheduling problems and

---

TABLE 4-1

Examples of Construction Activities That May Require Coordination between the Construction Team and the Operations and Maintenance Departments[a]

Construction Activity	Operations Involvement	Maintenance Involvement
1. Site excavation.	Participatory	Participatory
2. Equipment foundations.	Supportive	Participatory
3. Piping supports.	Supportive	Participatory
4. Pneumatic devices and tubing.	Supportive	Supportive
5. Electrical wire and cable tray.	Supportive	Supportive
6. Installation of control panels, distributed control systems.	Participatory	Supportive
7. Installation of all process connected electronic transmitters and sensor devices.	Supportive	Supportive
8. Installation of control valves and meter runs.	Minimal	Supportive
9. Installation of relief valves and rupture discs on vessels etc.	Minimal	Supportive
10. Interferences between pipe and conduit along pipe racks and building walls or ceilings.	Participatory	Supportive
11. Tie-ins to existing vessels, piping, and equipment.	Lead	Lead

[a]The levels of involvement shown in this table usually depend on whether it is a greenfield construction or construction in an existing facility. The reader needs to make a conscious decision of what the levels of involvement are.

construction incidents. When there is an unresolved concern or conflict, the project manager should be consulted. Every possible effort should be made to avoid the "us versus them" attitude.

Inadequate communication between the project team and the operations and maintenance departments can lead to serious incidents. For example, a construction project involved the installation of additional air dryer capacity in an air separation plant. To regenerate the additional capacity, a steam line was being installed for a new regeneration heater. Both the area foreman and the construction foreman knew about the work but miscommunicated about when the new line would be tied in to the steam system. The construction crew cut into an active high-purity oxygen line, which caused a serious fire. If the area foreman and construction foreman had been communicating properly, the incident could have been avoided. The operations and maintenance departments should have a lead role in authorizing a tie-in to an existing system. An operations or maintenance department representative should have been present at the site to point out the correct line and ensure that the work could be performed safely.

4.1.2. Control of Specific Construction-Related Activities

In addition to checking that construction is done according to the P&IDs and other specifications, the operations department should also be responsible for personnel and equipment safety, all process tie-ins, and all construction permits (i.e., work permits, hot work permits, welding permits). For example, if hot tapping is needed, the operations department will determine the appropriate time and specific conditions for the hot tap. Construction techniques, on the other hand, are usually monitored by the maintenance staff. For example, the rigging for lifting equipment above operating equipment should be carefully checked by the maintenance department. In this situation, the operations department would develop contingency plans for potentially hazardous incidents that could occur during the lifting of equipment. Table 4-2 presents some of the construction-related activities when operations and maintenance staff should have a certain degree of control.

4.1.3. Inspection of Equipment Installation

The inspection and testing of major equipment at the factory is critical and is usually carried out by the project team. However, the inspection of equipment as it is installed is also important to ensuring the overall integrity of the process and equipment. This inspection has a dual significance. First, the operations and maintenance staff help the project team with quality assurance and the verification of specifications. Second, the inspections allow operations and maintenance supervisors to gather the information needed to maintain process and equipment integrity during post construction activities. For example,

TABLE 4-2
Examples of Construction Activities Controlled by the
Operations and Maintenance Departments

Electrical work
General rules for the safe performance of electrical work are established by maintenance personnel. Electrical area classification should be known by all personnel performing electrical work while equipment is operating. Written authorization to proceed with work in operating areas should be obtained from the operations supervisor.

Pneumatic hammers and chisels
Maintenance department may have specific guidelines for using pneumatic hammers and chisels. If specific guidelines do not exist, operations and maintenance personnel should ensure that appropriate precautions are taken to avoid hazards associated with flying scale or chips and from tools and bits flying from the gun.

Hot tapping
Specific approval of the operations supervisor and appropriate permits should be obtained before hot tapping operations begin. The line to be hot tapped should be positively identified and the operations supervisor should mark the specific point on the line where the hot tap is to be made. The operations supervisor should also be satisfied that the metal thickness of the line is sufficient to prevent burning through during welding. If any alternate welding method is used, the maintenance supervisor should approve it.

Scaffolds
It is important that scaffolds be adequate for the intended use and that the workmen engaged in building and using scaffolds be thoroughly briefed in and about potential hazards within the area where the scaffold is to be used. Scaffolds should not block emergency controls or escape routes.

Hot work
The operations department should control all hot work permits for operating areas. Activities that might release hydrocarbons (e.g., sampling or gauging) in the area should not be performed unless approved by the individual who signed the hot work permit.

Confined space entry
Procedures for entering confined spaces are usually established by plant management. Operations should verify the conditions and issue permits for confined space entry.

the operations department requires information on the location and configuration of equipment, controls, and indicators on each piece of equipment to refine operating and training procedures. The maintenance department requires the knowledge of assembly procedures, valve and packing types, gasket material, instrumentation (from maintenance viewpoint), manufacturers model number, and spare part requirements to develop maintenance procedures. Table 4-3 lists some major pieces of equipment and includes items that should be inspected by operations and maintenance staff during construction.

TABLE 4-3
Operations and Maintenance Items of Interest for Major Equipment
Fabrication, Construction, and Installation

Major Equipment	Operations and Maintenance Items Of Interest
Boilers	Fuel gas knockout drum Flameout protection Blowdown Instrumentation (firing control, water level control, pressure) Safety valves Block valve locations Feed water systems and treating
Compressors and blowers	Lubrication Automatic shutdowns Emergency shutdown Surge control Vibration monitor Indicating instruments Crankcase vent Grouting
Drivers (steam turbine, gas engine, gas turbine)	Materials—cylinder, valves, wheels, case, shaft Packing vent Lubrication system Overspeed trip Automatic shutdowns Instrumentation Vibration monitor Safety valves Crankcase vent Grouting
Electric Motors	Electrical classification Proper grounding Lubrication Breakers and heaters
Electrical distribution system	Ground detector Area classifications and equipment use Grounding Sealing Vulnerability to fire Uninterruptible power supply system, emergency electrical system Communication systems Motor control centers, identification of starters

TABLE 4-3 *(Continued)*
Operations and Maintenance Items of Interest for Major Equipment
Fabrication, Construction, and Installation

Major Equipment	Operations and Maintenance Items Of Interest
Fired heaters and furnaces	Flameout protection Fuel gas scrubbers Snuffing and purging steam Curb walls and drainage Fireproofing of supports Tube metal temperature indicators Remote shutdown—fuel gas and process lines Flame pattern
Heat exchangers	Materials Layout, congestion, and exposure Maintainability (room for cleaning and maintenance) Leakage and tube failure protection Thermal expansion

Instrumentation and wiring	Air failure considerations (air open or air close)
	Power failure
	    Emergency power source
	    Valve action (close or open)
	Emergency control
	Suitability of instruments for service
	Cable tray; proper bonding and wiring
	Calibration and tuning
	Condition after construction activities
	Winterizing

Loading racks	Loading rates (open dome) Grounding, bonding, and insulation Downspouts (open dome) Loading connections Vent and drain facilities Emergency shutoff
Pressure vessels	ASME rating suitability Emergency relief valves Venting and draining Safe termination of emergency relief discharges Blinding Support fireproofing Small pipe and gauge connections Winterizing drawoffs

Continued on page 66

TABLE 4-3 *(Continued)*
Operations and Maintenance Items of Interest for Major Equipment
Fabrication, Construction, and Installation

Major Equipment	Operations and Maintenance Items Of Interest
Pumps	Materials Layout, congestion, and exposure Mechanical seals Vibration Drains and vents Instrumentation and controls Valving arrangements Sparing of pumps

Tankage	Atmospheric cone roof
	    Vent capacity and types
	    Weakened roof seam
	    Equipment in dike
	    Water draw-off
	    Overfilling protection
	    Spacing (between tanks, other equipment, property line)
	Floating roof
	    Control of roof drainage
	    Valved piping material inside dike
	Miscellaneous drums and small tankage
	    Venting capacity
	    Fireproofing of supports
	Pressurized storage
	    Safety valve arrangement and impingement
	    Fireproofing of supports
	    Fireproofing of vessels or water deluge
	    Small pipe connections
	    Gauge glasses
	Vacuum breakers
	Proper inert gas blanketing

4.2. MATERIALS OF CONSTRUCTION

Mechanical design restrictions are largely determined by the materials of construction. Materials are chosen for their ease of fabrication, ability to withstand service conditions, compatibility with other construction and process materials, and relative capital and maintenance costs. Using materials other than the specified materials of construction can create major hazards. In addition, documentation about the proper materials of construction is also essential for future reference and plant changes.

An improper use of materials of construction can lead to incidents that damage equipment and cause potential injury. In a plant that was producing hydrogen by the partial oxidation of crude oil, a control valve on the reactor outlet was replaced. The new valve was a different grade of steel than had been originally specified. After three months of service, hydrogen embrittlement caused the valve body to fracture during operation. A major fire and significant damage to equipment resulted. This incident could probably have been avoided by the proper control of the use of materials of construction.

The operations and maintenance departments should be satisfied that the materials of construction used are those specified, even to the point of verifying the materials of construction in some cases. The construction team may conduct tests as specified in the construction bid, to determine the properties of materials. First- and second-line supervisors should ensure that proper documentation is kept on the materials of construction, including test results. Quality assurance programs, in some cases required by governmental regulations,[1] can probably be managed most effectively by the maintenance department.

A thorough inspection of equipment by operations and maintenance personnel during plant construction is one of the most important steps in ensuring a safe startup and long-term process and equipment integrity. The operations and maintenance staff provide critical support in verifying that construction is performed as specified by design (e.g., leveling weirs during the construction of distillation columns). The materials of construction documentation that should be collected includes mill certification reports and verifications that the correct materials have been used, as well as any results of tests done on the materials.

4.3. CUSTOM EQUIPMENT FABRICATION AND INSPECTION

Custom equipment fabrication requires additional documentation such as a materials list, a parts list, parts drawing, and shop assembly drawing. The thorough inspection of custom equipment is especially important to ensure adherence to design specifications. When custom-built equipment is involved, operations and maintenance representatives should periodically observe the fabrication and assembly. This will enable the operations staff to understand the equipment and incorporate this knowledge into appropriate operating procedures, and the maintenance staff will be better able to address assembly and repair procedures. For example, equipment with special internal configurations should be inspected in the shop; this includes large compressors, large steam turbines, specialized reactors, and other custom equipment. It can be

---

1 Federal regulations [*29 CFR 1910.119(j)*] in the United States requires the implementation of a quality assurance program to ensure that equipment as it is fabricated is suitable for the process application for which it will be used. Appropriate checks and inspections must be performed to ensure that the equipment is installed properly and is consistent with design specifications. Also, the quality assurance program must ensure that maintenance materials, spare parts, and equipment are suitable for the process application for which they are used.

beneficial for operating and maintenance personnel to attend shop inspections and tests of critical equipment, for example, the inspection of shop-installed column or vessel internals.

An inspection of critical equipment during construction can identify faulty fabrication or installation. For example, during an assembly inspection of a multistage cryogenic pump, a maintenance foreman observed that the vanes on the impeller were opposite the normal rotation. Subsequent investigation revealed that the impellers had been cast in reverse because of an error in duplicating a shop drawing. This case demonstrates the importance of involving the operations and maintenance departments in the inspection process.

4.4. FIELD INSTALLATION

Pre-startup and commissioning activities usually involve the final inspection, testing, and commissioning of all the piping, vessels, equipment, and controls. However, many problems can be avoided by carrying out construction inspections that are focused on overall plant safety, operability, maintainability, and ergonomics. Operations and maintenance personnel can verify that the equipment is installed according to the specifications and manufacturer's instructions. They can make sure that the piping and instruments are installed according to the drawings. Table 4-4 is a checklist of items that the operations and maintenance departments may inspect while construction is in progress. In addition to piping, the installation of the protection systems should be monitored. As soon as a startup team is formed, the operations and maintenance personnel that have been performing construction inspections should be assigned to the startup team, which will ensure continuity of the inspection process and provide the startup team with people already familiar with the process and equipment.

4.4.1. Piping Installation

Surveys show that the failure of piping is a frequent cause of process plant accidents. According to Kletz[1], most piping system failures are caused by the piping not being installed according to design specifications or according to good engineering practice, particularly when details are not specified. The prevention of such failures requires more detailed specifications and a thorough inspection during construction. Operations and maintenance staff can play a lead role in spotting improper piping installation. Table 4-5 is a list of piping construction and installation items that should be checked.

Significant hazards are sometimes caused by seemingly unimportant piping details. For example, in the storage of a low-boiling hydrocarbon[2], only large gate valves were installed since it was believed no water would get

---

TABLE 4-4
Examples of Operations and Maintenance Inspection Items
during Construction

---

1. Make sure all instrument items are accessible and visible. This should be determined during the design phase. However, design errors can occur, so this should be checked accordingly.

2. Avoid locating instruments in areas of high personnel traffic.

3. Turn pressure gauges and dial thermometers toward traffic lanes for easy observation by operators.

4. Cover glass fronts on instrument items where heavy traffic or work might cause breakage.

5. Make sure glass and identification plates are covered before painting is done in the vicinity of instruments.

6. Make sure all necessary calibration (i.e., relief valves, pressure transmitters, etc.) is completed before installation unless in-place calibration is required.

7. Do not install in-line devices (e.g., orifice plates, positive displacements meters, rotameters, etc.) until lines have been flushed.

8. Do not run pneumatic tubing too close to hot surfaces.

9. Do not locate instrument and control tubing where they will be stepped on or otherwise damaged.

10. Make sure that electrical installations comply with the classifications.

11. Local storage of safety apparatus (e.g., breathing apparatus), essential ancillaries and tools (e.g., ladders and steps), solvents and cleaning materials, as well as waste materials (until disposal), should be provided.

12. The space needed for maintenance operations should be taken into account, and checked that it does not obstruct fire exits or encroach on busy passageways.

13. Fixed ladders or stairs should be provided to give access to overhead workpoints that require the presence of an operator or mechanic.

14. All electrical cables and apparatus should be protected against mechanical damage, heat, water, chemical attack, and unauthorized access.

15. A suitable attachment point (beam, eyebolt, etc.) should be provided above the top manhole or other opening in vessels, hoppers, etc. to enable a chain block to be used to get equipment in or out or to winch out an injured person.

16. Vessels, heat exchangers, etc. should be provided with a means of positive and complete isolation (e.g., blind flanges) so that they can be completely isolated during internal inspection and maintenance.

17. All areas containing flammable liquids should be electrically bonded and grounded, including steel-framed buildings.

18. Major units should be clean and free from obstruction.

19. Process contact surfaces should be as specified (correct materials of construction).

20. Equipment should be leaktight to specified test pressures (particular attention should be given to mechanical seals).

21. The tightness and flow of steam tracing should be checked.

22. Strategic lagging requirements should be checked.

23. Valve and instrument labels should be clear, visible, and legible.

TABLE 4-5

Example Piping Construction and Installation Checklist

1. Low- and high-pressure alarms should be installed in piping, as appropriate.

2. Piperacks should have an accepted level of fire protection and, if necessary, spray protection for piping in racks.

3. Piping exposed to corrosive atmospheres should be painted or otherwise protected to prevent the deterioration of wall thickness.

4. Plant or emergency personnel should be able to isolate piping by closing valves.

5. Valves should be clearly marked. In some cases, a comprehensive color coding system of pipes and valves may be useful.

6. Loops such as circle bends, double offsets, and other geometries involving complex geometry should be avoided.

7. Appropriate pipe lug supports, pipe restraints, guides, and anchors should be used.

8. Piping installation should match P&IDs and isometric drawings.

9. Are there necessary sample connections for routine samples or stream analyzers?

10. Low-point drains should be positioned in places so that the maximum amount of water can be drained after hydrostatic testing or for future maintenance.

11. Verification of the use of appropriate materials of construction should be made.

12. Are there provisions to accommodate expansion and contraction in pipelines?

13. Pipings should be labeled clearly to indicate different materials.

---

into the system. Because of flow reductions, however, water was suspected and an operator was sent to check the system. When he opened the 2-inch drain valve, nothing came out at first, so he opened the valve wider. Finally, the hydrocarbon started flowing out. When he tried to shut the valve, it stuck, probably from hydrates formed as a result of the refrigeration caused by the expansion across the drain valve. The operator was overcome by the hydrocarbon vapors. There had been recent rains and the dike area contained about six inches of water. The operator fell into the water and drowned. Of course, there were a lot of safety errors in this case, but a smaller drain valve and contingency procedures to cope with hydrate formation could have prevented a tragedy.

4.4.2. Pressure-Relief/Vent Collection

Verifying the correct installation of pressure-relief devices (i.e., control valve, spring or pilot operated relief valves, ruptures discs) is a necessary part of construction inspection. If process or equipment integrity is compromised, the favored mitigation action is actuation of the pressure indicator control valves and venting; however, in certain circumstances, the relief valve or the rupture disc may release to the atmosphere. It is very important that all pressure-relieving devices are installed properly, with the shipping packing removed from the conservation vents. Some key issues to remember about the installation of pressure-relieving devices are:

- In fouling service, sometimes rupture discs are installed below relief valves to prevent fouling and interference with the relief valve's operation. Set points of the rupture disc and the relief valve should be compared for compatibility.
- For pilot-operated relief valves, the sensor tubing should run to the ground. This allows for easy testing and inspection of the relief valve.
- Block valves (if used) for all pressure-relieving devices should be sealed open.
- The trip point of the relief valve should be indicated on the name tag.
- The specifications on relief device fittings should be checked very carefully. Figure 4-1 shows the appropriate specifications on fittings on a rupture disc fitted on top of a vessel. A common mistake sometimes found is the installation of a 150 ANSI flange at A instead of the proper 600 ANSI flange.
- To gain some redundant safety advantage, relief valves and rupture discs are quite frequently used in parallel. In those cases, the installation should be as shown in Figure 4-2.
- All pressure-relieving devices are inspected by the maintenance shops before construction can proceed with installation. There are variations of this procedure; however, in most cases, the maintenance department plays a lead role in inspecting and validating pressure-relieving devices.

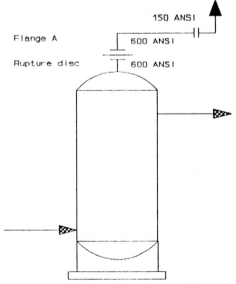

FIGURE 4-1. Installation of rupture disc.

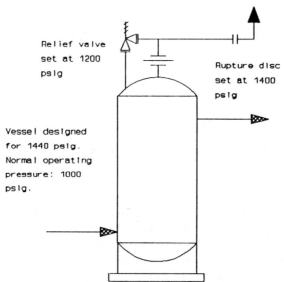

FIGURE 4-2. Installation of relief valve and rupture disc in parallel

4.4.3. Other Safety Systems

Depending on the process plant, the nature of operations, the severity of operating conditions, and the surrounding environment, various safety systems may be installed. Such safety systems may consist of protection, detection, or actuation devices. Examples of such systems are fire, gas, and smoke detectors, fire protection equipment, and emergency utilities. Although many of these safety systems are common to a number of plant sites, the actual design of the equipment is site specific. Operations and maintenance personnel should witness the on-site testing of safety protective systems to become familiar with their operation.

4.5. EQUIPMENT RECORDKEEPING

Equipment recordkeeping is essential for long-term process and equipment integrity. Also, some government regulations require that mechanical integrity and equipment records be maintained at all times.[2] When construction begins, the maintenance department should start accumulating data for the equipment files. All the information collected should be filed in a separate folder. Ideally, each folder should include at least the information shown in

---

2 In the United States, Department of Labor regulations[3] *[29 CFR 1910.119(j)]* contain specific guidelines about mechanical integrity that covers equipment recordkeeping. A summary of the U.S. regulations appears in Appendix A.

Table 4-6. These files, along with the other information and data pertinent to the process and its equipment, should be stored in a location that will be accessible day and night to all operations and maintenance personnel.

Integrated document management systems are a new trend in the chemical industry. A central database containing all plant documentation, from process drawings to construction memoranda, is the goal of modern computer-based document management systems. A wide variety of these systems are currently available, from the simple system for a small plant to a local area network for larger facilities. Vendors are using CD-ROM devices to provide detailed construction and maintenance manuals with graphics and sometimes multimedia video tutorials. Regardless of the data management systems used, the operations and maintenance departments should have knowledge of and continuous access to the data. Without knowledge of and access to the equipment data, it is impossible for the operations and maintenance departments to ensure the continued integrity of the plant processes and equipment.

When a wide variety of plant documentation systems are available, two things are necessary for successful information management:

- The information should be widely available in an easy-to-use and understandable format. Hence, training operations and maintenance staff in document access is critical.
- The authorization to change the master document in the system should also be tightly controlled and a system to inform many users of changes is also very important.

TABLE 4-6
Minimum Data Needs for Equipment Recordkeeping

1. Assembly drawings.
2. Parts lists.
3. Erection and maintenance manuals.
4. Catalogs.
5. Correspondence between the manufacturer and the project team.
6. Specifications
   (a) Specification sheet with all the name-plate data.
   (b) Material specification of wearing parts.
   (c) Part numbers and specifications.
7. Bulletins prescribing correct methods of performing unusual or difficult repairs.
8. Catalogs listing interchangeable parts.

4.6. SUMMARY

The role of operations and maintenance departments in the execution of specific activities during the various phases of the life cycle of a plant may range from a minimal to a lead role, as explained in Chapter 2. However, as the plant goes through the design phase, the construction phase and on into the pre-startup and commissioning phases, the overall participation of operations and maintenance departments continues to increase. The overall participation of the operations and maintenance departments during the design phase is minimal, although there may be certain specific activities of the design phase in which operations and maintenance departments have a more significant role. The degree of participation of these departments remains constant during the construction phase and consists mainly of oversight, assistance with inspections, and gathering data for operating procedures and mechanical integrity programs.

4.7. REFERENCES

1. Kletz, T.A. "Safety Aspects of Pressurized Systems." *IMechE,* 1981.
2. Jenett, E. "Safety in Process and Mechanical Design." *Chemical Engineering,* Aug. 17, 1964.
3. *Process Safety Management of Highly Hazardous Chemicals; Explosives and Blasting Agents; Final Rule, 29 CFR Part 1910.* Department of Labor, Occupational Safety and Health Administration, Washington, DC, February 24, 1992. *Federal Register,* Vol. 57, No. 36, pp. 6356–6417.

5

PRE-STARTUP AND COMMISSIONING

Planning and preparing for pre-startup and commissioning are the keys to safe, successful plant startups. Operations and maintenance personnel develop detailed plans and schedules of tasks to be performed during this phase. In the pre-startup and commissioning phase of a plant, operations and maintenance staff roles change from participatory to lead roles. This chapter describes the roles of the operations and maintenance departments when preparing for startup during the pre-startup and commissioning period.

Pre-startup activities cover the completion of the detailed design of the plant to the beginning of startup. During this period, a startup team prepares the startup, operating, and shutdown procedures; hires and trains operations and maintenance personnel; prepares detailed schedules; and performs a pre-startup safety review. The startup team establishes the safety procedures to be used by the operations and maintenance departments when performing the field inspections and commissioning equipment. After the construction organization declares that construction is mechanically complete, the operations and maintenance departments perform field inspections to confirm that the construction conforms to the design. After the field inspections, the equipment is commissioned in preparation for startup.

Commissioning the equipment is the functional check of the process up to the point when actual process materials are introduced. Commissioning activities include: pressure testing, cleaning, flushing, drying, purging the piping and vessels, utility checks, and process equipment testing. Vibration measurements, hot alignment of machinery, chemical cleaning, and control system checkouts should be completed during this period.

All field inspections and commissioning activities are essential to the safe initial startup of a process unit. During the pre-startup period, a pre-startup safety review should be performed.[1] The operability and reliability of the safety interlock systems, the alarms, and the safety relief valves should be established beyond any reasonable doubt during this period. How well operations personnel understand the process, startup and shutdown procedures, the fire protection system, and the emergency response plan are important considerations for a safe plant startup.

5.1. ORGANIZATION AND ROLES

The organization of the startup team and the roles of the operations and maintenance departments during pre-startup will depend on the method of construction. For a turnkey project, the contractor usually prepares the startup, operating, and shutdown procedures and does the inspection and commissioning. In some turnkey projects, the contractor trains the operators before turning the plant over to operations after the plant meets the specified production rates.

For non-turnkey projects, a common arrangement is for the contractor to be responsible for making certain that the mechanical equipment is in proper working condition and that the drivers and rotating equipment have been aligned. The contractor is usually also responsible for cleaning out and hydrostatically testing vessels, tanks, and piping. He should be available to the startup team for information and advice. The organization of the startup team and the roles of the operations and maintenance departments described in the following subsections are for non-turnkey projects. Regardless of the organization, an important role of the operations and maintenance supervisors is to assure that the contractor's performance is audited.

5.1.1. Startup Team

The nature of the project determines the selection of the startup team. Members should be selected according to their background and expertise in carrying out the specific assignments of pre-startup, commissioning, and startup. The startup team should consist of the following types of personnel: (1) technical operating specialists consisting primarily of engineers selected especially for startup preparations; (2) a plant operations group that maintains supervisory and line control over the operators and that will assume technical control of the plant after startup; (3) a maintenance group that may be part of the normal plant staff but that will be supplemented with additional technicians during pre-startup, commissioning, and startup; (4) a laboratory group

1 In the United States, Department of Labor regulations[4] [29 CFR 1910.119(i)] contain specific requirements for pre-startup safety reviews. A summary of the U.S. regulations is given in Appendix A.

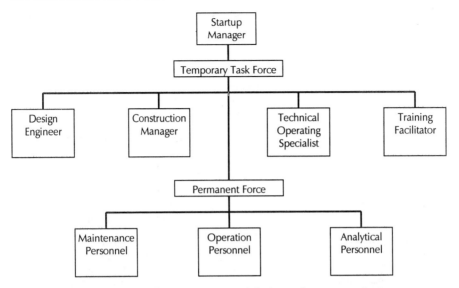

FIGURE 5-1. Example pre-startup, commissioning, and startup organization.

that will be part of the normal plant staff but that will provide additional technical support during the startup period.

Besides these core groups, the construction group should be part of the startup team. During the pre-startup period, the correction of any construction deficiencies will require cooperation between the construction group and the startup team to avoid time-consuming interferences. To provide project continuity, a design engineer who has been with the project from the conceptual design stage is an important member of the startup team. Figure 5-1 shows an example startup team organization. A training facilitator is included for a large project using a new process that would require extensive training.

The core groups defined here should work together as a team. The lines of communication, responsibility, and authority should be established early, and the startup manager should be firmly in charge. Once pre-startup and commissioning are underway, the startup manager should have technical control of the plant, and operations and maintenance employees should be supervised through the proper department channels. Control of the construction work will depend on the arrangement with the contractor. The groups should agree about all the details and, for this purpose, a startup committee composed of the heads of the various groups should meet daily or semi-weekly. Table 5-1 shows the groups and their membership that may be part of the startup team[1]. Also shown in the table are the members that should be included on the startup committee. Although all individual personnel shown in Table 5-1 may not exist, each of their functions should be addressed.

TABLE 5-1
Example Startup Team Organization

Startup Committee
Startup manager
Design engineer
Plus all personnel below with *

Technical Operations Group
Chief operating engineer*
Shift engineers (4)
Specialist engineers
Process
Instruments
Mechanical equipment

Analytical Department
Chemist*
Specialists
Chromatograph
Special analyses
Shift analytical technician

Construction Group
Construction manager*

Plant Production Department
Area supervisor*
General foreman
Shift foreman (4)
Computer and board operators
Field operators
Storage and loading men

Maintenance Department
Maintenance engineer*
Instrument engineer
Instrument technicians
Instrument fitters
Electrical and power engineer
Electrician
Pressure vessel and piping engineer
Pipe fitters
Welders
Insulators
Riggers
Draftsmen
Machine engineer
Machine fitters
Cleanup labor

5.1.2 Role of Operations and Maintenance Departments

The primary process safety roles of the operations and maintenance departments are to verify that the plant is built to specifications; write startup, operating, and shutdown procedures; train operations and maintenance personnel; develop a maintenance program; participate in a pre-startup safety review; and commission the equipment. Most tasks that concern the inspection and commissioning of the plant are shared by the operations and maintenance departments. Since close cooperation is needed among the operations department, the maintenance department, and the rest of the startup team, the exact division of responsibilities and duties should be defined in a formal startup plan.

The operations department is usually responsible for safety procedures such as lockout/tagout, confined space entry, and hot work permits. The operations department writes the startup, operating, and shutdown procedures. Additional people should be trained in basic skills, safety, and unit operations. New and experienced operators are then trained in the operations of the specific plant. Along with simulating plant operations, the operators become familiar with the plant by performing some inspection checks such as whether the piping has been completed, operability, accessibility, provisions for sampling, and critical equipment details.

The maintenance department is responsible for training its staff in setting up shop, and equipment; inventorying spare parts, lubricants, and materials; special tools and procedures; preparing routine maintenance schedules; and establishing equipment inspection procedures. Some inspection tasks performed by maintenance staff are checking safety valve settings, piping expansion and support, winterizing details, the cleanliness of critical piping, equipment expansion, insulation, heat tracing, packings, and lubricants. Most pressure testing and cleaning of equipment and piping is done by maintenance staff. During the commissioning of utilities and equipment, the maintenance department performs most of the tasks in close cooperation with operations personnel. During the final preparation for startup, the maintenance department performs such tasks as calibrating instruments, installing orifice plates, inspection and testing, vacuum and pressure, charging the catalysts, and providing temporary piping.

5.2. PLANNING

Involving operations and maintenance personnel in planning will increase the process safety in the plant. Planning is the key to a safe and successful plant startup. During the pre-startup period, sections of the plant should be opened to the operating group when they are mechanically complete. Thus, inspection, testing, and commissioning can be completed as early as possible. A detailed schedule is essential to coordinating work with construction and should be prepared by operations and maintenance personnel when the schedule can be coordinated with representatives of the startup team. The detailed schedule should be part of a written plan that covers the following items:

- Information center
- Inspection and commissioning
- Operating preparations
- Maintenance program
- Laboratory analysis
- Safety

A written plan can define the responsibilities included in each task to be performed during the pre-startup and commissioning period. It also ensures that communication channels are defined to minimize misunderstandings that could cause accidents.

An information center should be accessible to all members of the startup team on a twenty-four hour basis. The center may range from a simple filing system to a computerized information storage and retrieval system. For planning pre-startup and commissioning, information should be organized and placed in the center when the project reaches the end of the design phase. The following information should be available initially:

- Design basis
- Detailed process description
- Process flow diagrams (PFD)
- Piping and instrument diagrams (P&ID)
- Electrical drawings and classification areas
- Plot plans
- Process chemical safety data
- Safety procedures
- Detailed equipment specifications
- Utility use
- Vendor's drawings, operating instructions, and correspondence
- Relief system and design basis
- Ventilation system design
- Material and energy balances
- Safety systems (interlocks, alarms, detection, and suppression)
- Engineering design standards
- Fire protection system
- Process hazard reviews

As pre-startup and commissioning are completed, the following information should be added:

- Startup, operating, and shutdown procedures
- Inspection, tagging, and commissioning procedures
- Inspection and commissioning records
- Operator and maintenance training data
- Routine maintenance schedules
- Instrument and alarm settings
- Analytical procedures
- Calculation procedure for plant performance
- Emergency response plan

The inspection and commissioning of piping, vessels, and equipment require most of the effort during the pre-startup period; therefore, the detailed planning of each task is important. Using a checklist similar to Table 5-2, a detailed schedule should be developed to assign tasks for inspection and commissioning. A method should also be developed for tracking progress and making corrections. An example of a schedule is shown in Figure 5-2.[1]

TABLE 5-2
General Startup Preparation Checklist

Inspection
Vessel interiors
Vessel packings
Piping according to piping and instrument diagrams
Piping expansion and support (check free movement)
Equipment arrangement for access and operation
Cleanliness of critical piping
Insulation, steam tracing, etc.
Temporary strainers and blinds installed
Provisions for sampling
Electrical classification
Diking, curbing, and drainage

Pressure Testing, Cleaning, Flushing, Drying, and Purging
Pressure test piping and equipment
Flush and clean piping and equipment
Drain water to prevent freezing
Blow out piping
Orifice plates (checking bore and location)
Continuity testing with air
Dry out process
Chemical Cleaning
    Activation
    Passivation
    Rinse
Purge

Commissioning Utilities
Electric Power and Lighting
    Continuity check (megger)
    Trip settings at substations
    Isolation and safety
    Sample and check transformer oil
Water Treating
    Load filter beds
    Load ion exchanger
    Make up injection systems
Cooling Water
    Flush inlet headers, laterals, and return lines
    Drain to prevent freezing
    Clean tower basin
    Adjust tower fans
Service Air
    Clean air header by blowing
    Keep water drained
    Load desiccants and dry out header
Underground Drains
    Cleanliness and tightness
    Seals established
Steam
    Line-warming procedures
    Blow main headers
    Blow laterals

Condensate
    Disposal to drains initially
    Check trap operation
Inert Gas
    Identify and provide warnings
    Blow lines with air
    Isolate and purge, as required
Fuel Gas
    Identify and provide warnings
    Blow lines with air
    Isolate and purge
Fire water
    Flush and clean mains
    Check flow rate and pressure

Commissioning Equipment
Fired Heaters
    Instruments and controls checked
    Refractory dried out
Piping Strains on Equipment
Electric Motors
    Rotation
    Drying out
    No-load tests
Steam-Turbine Drivers
    Auxiliary lubrication and cooling systems checked
    Instrumentation and speed control checked
    No-load tests
    Light-load tests
Gas-Engine Drivers
    Auxiliary lubrication and cooling systems checked
    Instruments
    Idling tests
Gas-Turbine Drivers
    Lubricating oil governor and seal oil systems
cleaned
    Instrumentation and speed control checked
    Auxiliary heat recovery system
Centrifugal Compressors
    Lubricating and seal oil systems cleaned
    Instrumentation and controls checked
    Preliminary operation of lubricating and seal oil
      systems
    Operation with air
Vacuum Equipment
    Alignment rin-in adjustment
Pumps
    Alignment run-in adjustment
Hot-Alignment of Machinery
Vibration Measurements
Instruments
    Blow with clean air
    Dry out
    Check continuity, zero and adjust
    Calibrate as required

FOR: _____   PROJECT NO: ___9994___   PAGE NO: _____
PROJECT: _____   BY: ___AV___   DATE: _____
SUBJECT: PRE-OPERATIONAL SCHEDULE _____   ITEM NO: _____
            (ASSUMING A 4-MAN TEAM)

Task	Day 1 2 3 4	5 6 7 8	9 10 11 12	13 14
1. ACTIVATE UTILITIES (HEADERS AND SUB-HEADERS)				
- INSTRUMENT AIR	├————————	———————┤		
- COOLING WATER	├———————	——————┤		
- INERT GAS	├—————	————┤		
- STEAM	├——	——┤		
- BOILER FEED WATER		├————┤		
2. SYSTEMATIC INSTRUMENTATION CHECKOUT	├————————			
3. STEAM TRACING CHECK	├—		—AS———	—INDIV
4. TEST AND FLOW CALIBRATE MAIN AIR COMPRESSOR	├–AS-SOON--AS–	–PRACTICAL—	┤	
5. TANK CALIBRATION	├—			
6. WATER FLUSH AND CIRCULATION OF SALT SYSTEM		├———┤		
7. DRY SALT SYSTEM		├—┤		
8. LOAD AND MELT FUSED SALT			├————┤	
9. DRY AND TEST STARTUP FURNACE			├———┤	
10. ACTIVATE M.P. STEAM DRUM AND RELATED EQUIPMENT			├———┤	
11. WARM UP REACTOR WITH AIR			├——	┤
12. CIRCULATE SALT				├—
13. REACTOR HOT TEST (12 HOURS AT 400°C)				
14. COOL REACTOR				
15. REMOVE REACTOR HEADS AND INSPECT				
16. MECHANICALLY CLEAN TUBES				
17. CATALYST LOADING, PRESSURE TESTING AND ADJUSTMENT				
18. DEACTIVATE REACTOR HEADS, PIPING				
19. REPLACE REACTOR HEADS AND PIPING				
20. INSPECT SCRUBBER AND REFINER COLUMNS		├———	——┤	
21. RUN SCRUBBER WITH WATER			├———┤	
22. VACUUM TEST REFINER				├—
23. COMMISSION TEMPERED WATER SYSTEM				
24. WATER BATCH REFINER				
25. RUN XYLENE, CRUDE, REFINED PUMPS WITH WATER				
26. RUN SIMULATED DEHYDRATION, CHARGE XYLENE, CHECK DECANTER OPERATION				
27. COMMISSION PROCESS WATER SYSTEM				
28. DRY TANKS				
29. FLOW CALIBRATE BENZENE FEED SYSTEM				
30. CHECK PASTILLIATOR OPERATION WITH PURCHASED MAN				
31. FINAL ADJUSTMENTS AND FIELD MODIFICATIONS				
32. REACTOR HEAT UP				
33. START				

FIGURE 5-2. Schedule fpr inspecting and commissioning a maleic anhydride plant.. (Source: Finlayson and Gans. *Chemical Engineering Progress.* Vol. 63, No. 12, December 1967.)

15 16 17 18 19 20 21 22 23 24 25 26 27 28 29 30 31 32 33 34 35 36 37 38 39 40 41 42 43 44 45 46 47 48

AL——SYSTEMS——-ARE—— COMMISSIONED

Operating preparations include planning the safety procedures. Operations staff should plan the number of operating personnel, the mix of experienced and new operators, and training. The main activity of the operations department should be to plan and produce the startup, operating, and shutdown procedures.

The first- and second-line supervisors of the maintenance departments, because of their position and experience, are in a good position to do the following:

- Staff and train personnel
- Set up and equip shop
- Warehouse spare parts and materials
- Prepare special tools and procedures
- Establish equipment inspection procedures
- Obtain packings and lubricants
- File vendor equipment instructions
- Write necessary maintenance procedures

The first- and second-line maintenance supervisors should also help plan and complete a detailed maintenance schedule for the plant by the time of startup. They should help plan for auxiliary equipment such as cranes and block and tackle that may be needed to maintain difficult-to-reach access equipment.

Laboratory analysis planning should cover staffing and equipment, plus procedures for selecting which streams to sample, the frequency of sampling, and analytical procedures and calculations. A sample retention policy should also be established.

Safety planning covers handling hazardous materials, fire safety, physical and mechanical safety procedures, and personal protection equipment. Table 5-3 is a checklist of safety items that should be included in the pre-startup planning. Safety should also be integrated into operations and maintenance personnel training.

5.3. PREPARATION FOR STARTUP

Preparing for startup or pre-startup and commissioning involves executing the planning discussed in the previous section. Before the construction group opens mechanically complete sections of the plant, the operations department should establish procedures for lockout/tagout, confined space entry, and hot work permits. Next, operations personnel should write startup, operating, and shutdown procedures, then they handle the staffing and training of operators. The maintenance department should first establish equipment inspection and testing procedures and schedules. Maintenance staff then develops a maintenance program that includes routine maintenance procedures and schedules;

TABLE 5-3
Safety Planning Checklist

Safety procedures for lockout/tagout, confined space entry, hot work permits and excavation
Fire brigade
Fire-fighting procedures
Fire suppression chemicals
Safety valve settings and installations
Alarms
Safety interlock system
Detection for gas, smoke, and fire
Pre-startup safety review
Emergency response plan
Protective clothing: goggles, face shields, hard hats, work gloves, rubber gloves, aprons, hoods, gas masks (with spare canisters), and self-contained breathing apparatus
First aid and medical assistance
First-aid kits, blankets stretchers, antidotes, and resuscitators
Bunker suits, axes, ladders, hand extinguishers, hoses, wrenches, and nozzles

inventorying spare parts, lubricants, and materials; and setting up shop and equipment. Additional staffing and training are usually necessary when preparing for startup and operation.

5.3.1. Staffing Operations and Maintenance Departments

The operations and maintenance departments should help determine the number and skills of people that will be needed to operate and maintain the plant. Figure 5-3 shows an example operation organization for pre-startup, commissioning, and startup. This organization, except for shift engineers, is the same as it is during normal operations. Figure 5-3 shows the shift engineers with line authority. In many companies, the shift engineers have staff responsibilities and serve as consultants so they can devote more time to technical problems. For the pre-startup, commissioning, and startup periods, specialized workers will often augment the staff required for normal maintenance. Table 5-4 provides examples of maintenance requirements during pre-startup, commissioning, and startup[1].

The first- and second-line maintenance supervisors should help develop a written job description for each position. The job description may be required by government regulation and is usually very useful for finding experienced personnel within the company and for advertising outside the company. Company policy may require that the search be conducted through the personnel department. The job description will thus help the personnel department conduct initial screening before selected candidates are interviewed by the process-area supervisors. For a safe pre-startup and commissioning, it is advantageous to have experienced personnel. It may help to select a target.

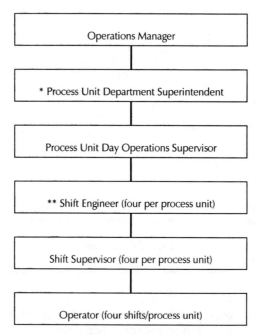

* member of startup committee

** for initial startup

FIGURE 5-3. Example Operation Organization for Pre-startup, Commissioning, and Startup.

TABLE 5-4

Examples of Maintenance Requirements for Specialized Workers for Pre-startup, Commissioning, and Startup

Craft	Normal Operation		Startup Dry Runs and Dynamic Tests
	Days	Shifts	
Machine fitters	1	—	3–6
Pipe fitters	2	—	up to 10
Pipe welders	2^a	1	up to 10
Electricians	1^a	1^a	1^a
Instrument technicians		1	1 + 2 on days
Instrument fitters	2	—	up to 4
Insulators	1^b	—	on contract
Riggers	2^a	—	up to 5
Piping draftsmen	2^a	—	2^a
Cleanup men	2	—	2
^{a}On call ^{b}For repair work only			

One suggestion is that no more than 60% of the work force consist of inexperienced employees. Hiring and training should begin at least four to six months before construction is completed.

5.3.2. Training

The proper training of operations and maintenance personnel is important for safe and efficient plant startup and operations. First- and second-line supervisors play a minimal to supportive role when determining the level and type of training required; however, they usually take a more significant role in ensuring that all personnel have completed the necessary training and are ready for specific assignments. Basic training for new employees should be given in areas of basic skills, basic safety, and unit operations. Besides the basic training of new employees, all operators should have specific job training, including at least:

- An overview of the process
- Operating procedures
- Specific safety and health hazards
- Emergency operations, including shutdown
- Safe work practices

Maintaining training documentation is mandated by government regulations and should be a part of the training program.

5.3.2.1. Basic Training

Basic training begins with training the trainers. When the basic trainers are operations and maintenance personnel, each department head should teach the first-line supervisors in the same way they wish these supervisors to teach their own personnel. The trainers are usually the first-line and second-line supervisors, or in some instances a training facilitator, as shown in the startup organization, Figure 5-1. Some companies use basic training materials supplied by outside organizations such as the Synthetic Organic Chemical Manufacturing Association (SOCMA) or the Nuclear Energy Council (NEC). An advantage of having basic training information already prepared is that it can be easily augmented, as needed.

The basic training should include:

- The job
- Basic science
- Basic safety
- Processes
- Utilities
- Equipment
- Fundamentals of measurement and instrumentation

The Job   Trainees should be taught that the operator's job is to convert raw materials into useful products through the safe and efficient control of plant equipment. The maintenance department's job is to keep equipment in top running condition by means of regular mechanical integrity and preventive maintenance procedures.

Science   The study of physical chemistry should include physical states, temperature, pressure, density, specific gravity, heat, vapor pressure, flammability, flash point, and combustion. Basic mathematics and chemistry should be covered, including topics such as units of measure, composition of matter, and properties of acids, bases, and salts.

Basic Safety   Trainees should be taught the potential hazards of handling materials, fire protection, physical and mechanical safety procedures, and personnel protective equipment. Trainees should be taught how to develop a plan to handle foreseeable emergencies.

Processes   Fluid flow and heat transfer should be taught, along with distillation, absorption, stripping, drying, cracking, gas compression, refrigeration, and steam generation, as appropriate to the processes of the plant.

Utilities   Trainees should be taught the use of air, gas, water, steam, and electricity. The availability, generation, and distribution of utilities throughout the plant should be discussed.

Equipment   The equipment information should include the functions and internal mechanisms of valves, pumps, steam traps, strainers, compressors, turbine motors, columns, heat exchangers and separators. The instruction should be augmented with models, cut-away models, equipment manufacturer's pamphlets, and field trips to observe the equipment in service.

Fundamentals of Measurement and Instruments   Measurements of pressure, temperature, fluid-flow, and levels should be explained. Controllers, motor valves, valve positioners, transmitters, and receivers should be covered, along with recorders with strip and circular charts, different types of scales and inking, and various drive mechanisms. Also the concept of computer process control should be introduced to the trainee.

     Table 5-5 lists a table of contents for basic training course material that can serve for both operators and maintenance personnel.[2] The material, prepared by SOCMA, covers basic skills, basic safety, applied unit operations, and an overview of U.S. government-regulated training requirements.

TABLE 5-5
Outline of Operator Training Manual

**Synthetic Organic Chemical Manufacturers Association
Chemical Process Operator Certification Training Manual Level 1**

Table of Contents

Introduction
Acknowledgements
A Chemical Process Operator Certification Program—Why?

I. Basic Skills

    A. Mathematics

    B. Chemical Concepts

    C. Physical Concepts

II. Basic Safety

    A. Hazardous Substances

    B. Fire Safety

    C. Physical and Mechanical Safety Procedures

        1. Confined Space Entry

        2. Electrical Safety

        3. Electrical/Valve Lockout/Tagout Procedures

        4. Cutting Welding and Burning

        5. Line Breaking

        6. Handling Compressed Gases

        7. Elevated Work Practices

        8. Hoisting Equipment and Its Uses

    D. Personal Protective Equipment

        1. Minimum Eye Protection

        2. Special Eye and Face Protection

        3. Safety Shoes and Special Foot Protection

        4. Hearing Protection

        5. Protective Clothing

        6. Respiratory Protection

III. Applied Unit Operations

    A. Reading Instrument Drawings

    B. Basic Instrumentation

    C. Utilities

    D. Piping

    E. Pumps

TABLE 5-5 *(Continued)*
Outline of Operator Training Manual

                    F. Valves

                    G. Fire & Emergency Response

                        1. Emergency Evacuation of Buildings

                        2. Fire Reporting and Investigation

                    H. Emergency Planning

Conclusion and Review

Overview of Federally Regulated Training

                    A. Respiratory Protection

                    B. Hazard Communication

                    C. Carcinogen Training

                    D. Asbestos Standards

                    E. Hazardous Waste Operations

                    F. Powered Fork Lift Truck Operation

                    G. Fire Systems

                    H. Portable Fire Extinguisher

                    I. Portable Fire Extinguisher Use

                    J. Hearing Conservation

References and Resource Materials

5.3.2.2. Specific Job Training for Operators

After new employees have received basic training, they are ready to join the experienced operators in specific job training. Since the operation of a unit requires specialized training and information, instruction should be conducted to acquaint operators with enough detail to perform their jobs safely. The training should be conducted after the operating manual has been prepared and should include:

- Introduction
- Outline of Process
- Operation
- Utilities
- Automatic Safety Equipment
- Safety
- Laboratory Analysis
- Maintenance

Introduction The introduction should acquaint the operators with the manual, the plant, and the process. It should define the operating organization, the chain of command, and the various operating jobs. The raw materials to be processed, by-products, intermediates, and products should be explained. Uses of the products and by-products of the plant should be included.

Outline of the Process Each unit operation in the plant should be described, along with each major piece of equipment and critical operating conditions. A process flow sheet encompassing all of the unit operations, major equipment, and primary piping should be reviewed to give a complete view of the process.

Operations The operations section should contain the entire startup, operating, and shutdown procedures for each unit operation, such as field storage, unit tank area, reaction, recovery, distillation, and compression. A complete instrumentation explanation should include the type of measurement, location, normal operating condition, and the high and low values of each sensor. The methods of controlling levels, flow rates, temperatures, and pressures should be explained, along with how the information is displayed-either by the computer or on control panels. Topics such as centrifugal pumps, agitators, mechanical seals, and miscellaneous operating duties should be covered.

Utilities Flow sheets showing the distribution of power and utilities in the plant should be presented. Alarms and instruction for the action to take immediately following a utility failure should be covered.

Automatic Safety Equipment A detailed explanation of the safety interlock system should be presented. The operators should know which valve will close or open with loss of power or air when the safety interlock system is automatically or manually activated. Also, the combinations of events that will trigger the various portions of the safety interlock system should be explained.

Safety Plant safety rules and special precautions for handling any of the process chemicals, along with their flammability, explosiveness, toxicity, and personnel exposure hazards should be reviewed, as well as Material Safety Data Sheet information. In case of a fire, isolation of the process area and procedures for extinguishing the fire should be explained. Preventing fires by eliminating fire hazards should be stressed. The emergency response plan should be reviewed so that each operator knows their responsibility.

Maintenance Preparing equipment for maintenance, repair, lubrication, gaskets, seals, and packing should be discussed thoroughly. The safety procedures for removing inventory, decontamination, lockout/tagout, confined space entry, hot work permits, and coordination with operations should be covered.

Operators should receive additional hands-on training by simulating the operations with water, air, or other suitable fluids during commissioning as the equipment is being flushed. Operator training can also be augmented by using a simulator in which upset scenarios are staged and levels, flow rates, pressures, and temperatures are programmed to change, depending on the responses of the operator. During the training episodes, the simulated plant operating variables and the response action of the trainee can be recorded for review purposes. With computer controlled facilities, simulations of the responses of the process variables can be programmed to react to the control settings to enhance operator training. Observation and participation in the actual operation of a similar facility or pilot plant are excellent training methods. Operations supervisors can take advantage of all the training opportunities during this period.

5.3.2.3. Specific Job Training for Maintenance Personnel

Maintenance personnel usually include pipe fitters, millwrights, mechanics, welders, electricians, and instrument maintenance technicians. Pipe fitters usually work in a group that maintains the pressure vessels and piping. As part of their basic training, they need to know about the packings of valves, pumps, stirrers, and mixers. Mechanics who work on engines should be given basic training on pumps and compressors. The basic skills training for welders consists of welding on carbon steel. Electricians should receive basic training in general wiring techniques for normal voltages, electrical breakers, and starters. Instrument maintenance technicians receive basic training in the calibration of transmitters for pressure sensors, thermocouples, and control valves.

After maintenance personnel have completed basic training, they should receive special training in new technologies, computer systems, and any hazards defined for maintenance of the plant. For example, mechanics and pipe fitters working on high-pressure systems may require special training. Mechanics may need special training in unusual types of turbines, compressors, and blowers. Some plants or processes may require the use of equipment or machines made from special alloys. Welders working in such plants would require special training in alloy welds and radiographing. Instrument technicians may need special training in vibration analyses for rotating equipment, and, depending on the training philosophy of the company, they may need training in instruments and sensors that interface with computers. All maintenance personnel may need some special or additional training if the plant is handling or processing unusual or special materials.

The special training requirements discussed here are only examples. The extent and breadth of the special training vary from plant to plant and are determined by the plant management on the basis of company guidelines, industry practice, and many other data sources. However, the first-line and second-line supervisors play a key role in helping plant management identify and provide the special training. To participate in plant inspections, personnel should complete training before the end of construction. Maintenance supervisors should take the lead in enuring that their personnel have every possible opportunity for training during this period.

5.3.2.4. Documentation of Training

The training of each operator and maintenance person should be documented. An effective way to document training is to develop procedures for certifying the completion of training, including training, testing, documentation, and continuing performance evaluation. A certification program will help ensure that workers in safety-critical jobs have the necessary knowledge and skills to perform their work safely and efficiently. Since they are responsible for training, the first-line and second-line supervisors play the lead role in documenting the training of all operations and maintenance personnel. An example of a certification developed by SOCMA is shown below.

The general program consists of four components:

1. *Level I—Basic Training*
   This training provides fundamental information considered to be essential to all chemical process operators.
2. *Level II—Process-Specific Training*
   This training provides process-specific training materials on individual unit operations which, when integrated into an individual plant's training program, can lead to process-specific operator certification.
3. *On-the-Job Training*
   This training consists of company-specific and site-specific training at the facility, with special emphasis on actual use of the equipment under close supervision.
4. *Demonstrations of Proficiency*
   This training consists of requiring trainees to demonstrate their ability to perform specific assigned tasks, for example, the startup or shutdown of a process, under close supervision.[2]

The documentation of training extends to contractors performing maintenance or repair, turnaround, major renovation, or specialty work on or next to the process area. The documentation should contain the identity of the contract employee, the date of training and the means used to verify the comprehension of the trainees.

5.3.3. Maintenance Activities during Pre-Startup

The first task of the maintenance department is to develop inspection and commissioning procedures. To ensure process safety it is imperative that the maintenance department develop safe procedures. When construction begins, maintenance should set up a complete equipment file. The list of equipment should include pressure vessels, storage tanks, process piping, relief and vent systems, fire protection, safety interlock systems, alarms, instruments, pumps, fired heaters, compressors, blowers, drivers, and control systems. For each piece of equipment, assembly drawings; parts lists; erection and maintenance manuals; catalogs; correspondence among the manufacturer, contractor, and startup manager; specification sheets with all name plate data, material specifications of wearing parts, and part numbers and specifications; bulletins describing correct methods of performing unusual or difficult repairs; and catalogs listing interchangeable parts should be located in the information center. Gathering and organizing this important information is essential to the safe and efficient maintenance of process equipment.

5.3.3.1. Maintenance Procedures

As detailed information becomes available during the pre-startup period, procedures for repairing critical equipment should be developed and any hazards associated with these repair should be noted. Any hazards that could develop from repetitive maintenance or repair should be carefully documented.

An example of the effect of a lack of procedures and inadequate training requirements on plant safety and loss prevention is described in Figure 5-4. The example illustrates the importance of establishing and giving training in maintenance procedures.

Factual description:
In an ethylene plant, regularly scheduled maintenance was being performed on a unit that consisted of a 12,000-hp steam turbine and a cracked gas compressor. The turbine was completely disassembled so that the shafts, bearings, turbine blades, seals, and interstage control valves could be inspected. During the maintenance, a pair of channel lock pliers was knocked off the turbine into the steam inlet line. This incident was compounded by the failure to conduct an adequate inspection before reassembly. Thus the pliers remained in the inlet nozzle of the turbine. After following normal warmup procedures using low-pressure steam to prepare the turbine for operation, the inlet valve for the 1,200-psia feed steam was opened to bring the machine to full speed. When the inlet valve opened, the pliers were carried into the turbine wheels destroying them. This resulted in the immediate cessation of startup activities; the plant was shut down and remained down for an additional two weeks.

Contributing causes	Process safety concept
1. Maintenance procedures and training should have ensured that an inspection of the equipment was performed before the job closed. If deficiencies are found during the inspection, supervisors should be informed and the equipment should not be brought on line.	Maintenance procedures and training
2. The checklists and other procedures used for the pre-startup safety review were not adequate.	Pre-startup safety review

FIGURE 5-4. Importance of maintenance procedures.

Using the equipment lists and the general startup preparation checklist shown in Table 5-2, the maintenance department should prepare a detailed checklist for each piece of equipment. Figures 5-5 and 5-6, respectively, contain examples of detailed check-lists for an air compressor and an instrument air dryer[3]

5.3.3.2. Mechanical Integrity

Ensuring the mechanical integrity of process equipment is an important element of process safety management. During pre-startup and commissioning, maintenance should develop a program for ensuring and maintaining equipment integrity. The program should include equipment inspections and tests and establish the frequency and criteria for acceptability and documentation of the inspections and tests.

In addition to the manufacturer's recommendations and company experience, the inspections, tests, and frequency should be guided by codes and standards. The program should develop regular external inspections of items such as foundations, anchor bolts, concrete or steel supports, guy wires, pipe hangers, grounding connections, protective coatings, insulation, vessel skirts, nozzles, sprinklers, and drains. Inspection and testing programs should be established for corrosion. Baseline data on material thickness should be ob-

FIELD INSPECTION CHECKLIST

For _____   Job No. _____

Location _____   Date _____

No.	Air Compressor	Yes	No	N/A	Remarks
	Before Compressor Is Started				
1.	Does installed pump meet specifications?				
2.	Piping of drainer packages on intercoolers properly located and installed?				
3.	Vendor expansion-joints properly installed?				
4.	Startup blinds removed?				
5.	Air filter properly installed, with screen drive operated and housing locked up?				
6.	Level alarms on drainer packages actuate at				
7.	Bearings adequately vented?				
8.	Interstage piping cleaned and inspected				
	After Compressor Is Started				
1.	Bearing temperature normal?				
2.	Joints soap-tested and tight?				
3.	Vibrations recorded daily during startup?				

FIGURE 5-5. Field inspection checklist for an air compressor.

FIELD INSPECTION CHECKLIST

For _____   Job No. _____
Location _____   Date _____

No.	Instrument-Air Dryer	Yes	No	N/A	Remarks
1.	Piping tie-ins ensure correct direction of air flow through the dryer? (As a minimum, the inlet piping to the dryer should include an oil filter.) Oil filter includes means for checking oil accumulation? (Oil drain valve?)				
2.	Vendor's internal piping and control equipment reviewed? Direction of air flow correct? Control equipment devices properly wired and piped?				
3.	Proper volume of desiccant loaded?				
4.	Upon availability of instrument air, check operation of transfer valves: Valves operate freely? On units with multiple valves operating in sequence are timer and sequence of operation of valves o.k.? Valve sequence ensures continuous flow of instrument air (i.e., standby power comes on the line prior to shutting down of operating tower)?				
5.	All safety devices (relief valves, automatic bypass valves, etc.) checked for proper operation and settings?				

FIGURE 5-6. Field inspection checklist for an instrument-air dryer.

tained during the pre-startup and commissioning period while vessel openings and critical inspection points are accessible.

Preventive maintenance schedules should be developed on the basis of equipment and instrument failure rates and on the company's experience. Documentation of the preventive maintenance results can be used for predictive maintenance. An example of predictive maintenance is estimating the time to a major overhaul of an internal combustion driver by analyzing the contamination of the used oil.

5.3.4. Development of Operating Procedures

The operations department is usually responsible for developing operating procedures. Some government regulations require that certain elements of process safety be contained in the operating procedures and that they be current and accessible.[2] The operating procedures include startup, operating,

2 In the United States, Department of Labor regulations[4] [29 CFR 1910.119(f)] contain specific requirements for the development and content of operating procedures. A summary of the U.S. regulations appears in Appendix A.

shutdown, and emergency procedures. The introduction of the operating procedures should include a list of critical variables and limits. The safety protective control systems should be described and include a tabulation of alarms and interlocks. For plants with computerized controls, the logic of the computer software and the data paths to the controllers should be described for the operators. The operating procedures developed during this period will set the tone for the future process safety of the plant.

The level of detail needed in operating procedures manuals varies from plant to plant. An example of how insufficient detail can affect safety is the Three Mile Island Nuclear Plant accident in 1978. The reactor cooling water was becoming overpressurized and a solenoid operated relief valve opened. The valve stuck open after the solenoid was de-energized. Unfortunately operators thought the control room indicator was directly measuring the valve position when it only measured the solenoid condition. Because they did not understand this indicated measurement, the operators did not properly respond to the continuing loss of cooling water.

5.3.4.1. Startup procedures

Procedures for initial, turnaround, partial, and restart startups should be addressed. The procedures should detail the step-by-step actions to be taken to bring the plant to full production. For the initial startup, the first step will be to complete a pre-startup checklist and introduce process fluid inventories into the units. The instructions should include the specific opening and closing of valves, any assistance needed from maintenance for temporary piping, which alarm and safety interlocks must be bypassed, and which process parameters are used for the basis of adjusting flow and recycle rates. Any potential process safety hazards should be explained. As the process parameters come into operating range, the procedures should explain how to activate the alarms and safety interlocks.

Startups after a turnaround or extended outage require the inspection and commissioning of any additional or repaired vessels, piping, or equipment. The startup procedures are the same as in the initial startup sequence. Partial startup procedures address the startup of a particular unit operating within a plant. It may be necessary to start up the particular unit when additional units are added to a process or when the temporary operation of the unit is a step in startup. An example of temporary operation is the charging of a distillation column with its operating fluid and operating the column on total reflux until the feed and output units are on line.

Restart startup procedures should address the temporary holding of a process in a safe state following a trip in the safety interlock system until the cause of the trip can be addressed. The restart procedures also define the steps to establish flow and bring variables back to normal operating ranges.

Figure 5-7(a) shows the controls for a packed distillation column[5]. The general pre-startup and startup steps shown in Figures 5-7(b) and 5-7(c) could

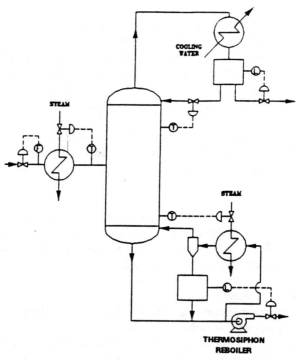

(a) Column control

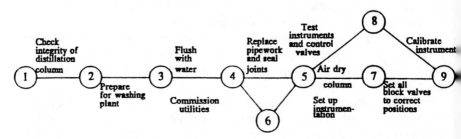

(b) Pre-Startup and commissioning steps

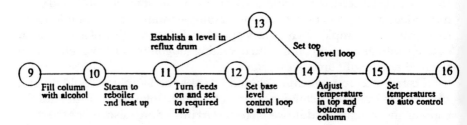

(c) Startup steps

FIGURE 5-7. Steps in the commissioning and startup of a packed distillation column.

be used to describe the overall process, followed by step-by-step tasks for executing the startup procedures. Following the detailed step-by-step startup procedure, an abbreviated checklist version of the startup procedures should be developed.

5.3.4.2. Operating Procedures

Once the process has reached operating conditions, the main consideration is maximizing the production of product that meets specifications. The normal operating procedures should define the target and range of the process variables for producing product with different specifications and for operating the plant at different turndown rates. Minimum stable turndown rates should be stated for each operating unit. The procedures should explain how operating parameters such as flow rate, temperature, and pressure affect the quality and rates of production. Step-by-step tasks should be defined for changing between operating states. The procedures should explain the consequences of operating the plant with critical parameters outside their specified ranges, as well as the steps to take to recover from upset conditions.

The procedures should specify the samples, analyses, and data sheets needed to monitor plant performance. Information to enter in the shift logs (such as any process upsets or maintenance items) should be explained, along with the checklists for inspections and checks of plant equipment to be performed each shift. The operating procedures should include instructions for performing tasks performed only as the need arises, such as the regeneration of dryers, decoking, and emergency power generation for instruments and control.

5.3.4.3. Shutdown Procedures

Procedures for normal, extended, and emergency shutdown should be written. Normal shutdowns are scheduled events that are usually determined by required inspection, preventive maintenance, equipment repairs, or process modifications. The normal shutdown may vary from the simple removal of a pump from service to a complex multi-week turnaround.

The normal shutdown procedure should give a detailed step-by-step narrative of tasks to take the process from an operational to a safe nonoperational state. The procedure should describe the steps for safely performing the removal of inventory, isolation, and the decontamination of each piece of equipment. For the extended shutdown procedures, additional step-by-step tasks should be written to provide for the long-term exclusion of process fluids from the vessels and equipment and for monitoring their status. Operators may initiate emergency shutdown because of a lack of containment, weather conditions, or threatening conditions from an adjacent process unit; because of equipment failure; or because of severe process upset conditions. For emergency shutdown, procedures should be developed for step-by-step tasks to maintain the process in a safe state until decisions can be made to resolve the emergency shutdown condition and restart the unit or proceed with steps for the removal of inventory, isolation, and decontamination. Potentially hazardous conditions should be noted for each of the shutdown tasks.

Operating procedures are usually contained in an operating manual, which typically has the following sections:

- Detailed Process Description
- Description of Utilities
- Process Equipment
- Preparation for Startup
- Operating Procedures
- Safety and Emergency Procedures

After a general organizational structure such as the one above is selected, the detail to be included in each section should be outlined. Examples for this stage of preparation are available in the literature. One such example[3] is reproduced here in Table 5-6 with minor modifications ("Section V: Safety" has been added to the original table).

Company and government safety requirements should be checked to ensure that all necessary requirements are addressed in the manual. Along

---

TABLE 5-6
Example Outline of Operating Instruction Manual

Title Page
Table of Contents

Introduction
This section discusses the content of the manual, briefly lists the type, purpose, and capacity of the plant, and describes the quality and quantity of the products.

SECTION I: THE PROCESS

1. Description of the Flow
Process Flows—This subsection describes the complete flow through the process system, making reference to vessels, instrumentation, pumps, and other equipment, but not detailing such data as line numbers or process conditions.
Auxiliary Flows—The auxiliary systems pertinent to the process are described in this subsection, such as ammonia refrigeration and the flushing of oil systems.
Utility Flows—Onsite units such as intricate steam-generating systems, air, or inert-gas systems are described in detail.
2. Process and Operating Principles
This subsection describes the theory and principles of the process, the process variables, and their effect on product quality. It also discusses the unit control, an understanding of which is necessary for efficient operation and quality results. Graphs and data may also be included for establishing operating variables.

SECTION II: CONDITIONING AND SPECIAL EQUIPMENT

1. Conditioning
This subsection covers conditioning of the equipment for startup and operation. It includes such subjects as breaking in pumps, compressors, and blowers; drying out furnaces and vessel linings; hydrostatic testing; inspection of vessels; flushing of lines; catalyst loading; checking systems auxiliary to the compressors; and dry runs.

2. Special Equipment
Mechanical Equipment—This subsection contains a detailed description of special or critical equipment such as compressors, blowers, large turbines, filters, special strainers, and reactor vessels.
Instrumentation—This subsection describes in detail complex instruments or circuits when elaboration is necessary for understanding and operation. Shutdown instrumentation is usually covered in this subsection.

SECTION III: STARTING PROCEDURE
This section describes the overall startup plan. It goes through process procedures step-by-step, in narrative form, giving details fully and noting special precautions. Unusual design considerations are then flagged for special attention. Catalyst activation is a typical procedure described in this section.

SECTION IV: SHUTDOWN PROCEDURE

1. Normal Shutdown
This subsection describes the overall shutdown plan used when the plant is being deliberately taken out of service.
2. Special Procedures
Step-by-step procedures and instructions are given for such auxiliary functions as catalyst regeneration and furnace decoking.

SECTION V: SAFETY

1. Chemical Properties and Hazards
Precautions are described to prevent exposure to hazardous chemicals, plus active and passive systems to contain a release, and MSDS information.

2. Emergency Procedures
Special emergency procedures and precautions are described in detail.

SECTION VI: APPENDIX
1. Standards
Standard instructions and descriptions, equipment lists, and tabulations are included in the first subsection of the appendix.
2. Drawings
This section contains drawings pertinent to the process and of interest to the operators: a plot plan, process flow sheets, an anticipated operating-conditions sheet, process piping and instrumentation flow sheets, auxiliary and utility piping and instrumentation flow sheets, vessel and furnace drawings, and data sheets.

with company policy statements and local regulatory requirements, the following items should be considered for inclusion in the operating manual.

- General safety instructions
- Management of change procedures
- Key regulations and reporting limits for releases
- List of support documents for future reference

Most of the information for the operating manual should be available in the information center described earlier in this chapter. The emergency response plan[6–8] will be needed for the emergency procedures part of the manual. The operations department may be responsible for writing the emergency response plan.

5.3.4.4. Emergency Response Plan

When a release of hazardous materials occurs and the active and passive mitigation systems fail to contain it, the emergency response plan should be activated[9,10]. An emergency plan includes both the response to incidental and more serious releases.

For minor releases, the emergency response plan should specify whether employees should evacuate to a safe area. If employees are to move to a safe area, a distinguishable method[11] should be available to alert all employees in the process area. The path that the workers take from the process area to the designated safe area may vary, thus a way to find evacuation routes should be established. When workers in the process unit are designated to control the emergency, the response plan should specify the makeup of the response team, protective equipment, and training.

For serious releases, the emergency response plan should cover the notification of workers and routes of safe egress. The plan should designate an incident commander and establish an emergency control center. The center should be equipped with the following:

- Communication center (between commander, company and local community)
- Backup communication systems (telephone and radio)
- Backup power
- Notification lists
- Mutual aid information
- Weather data sources
- Dispersion and explosion modeling results
- Emergency response equipment
- Primary and alternate escape routes
- Material Safety Data Sheets
- Emergency plans and procedures

- Plant drawings (plant layout, fire protection system and emergency lighting)
- Company responder resources

The plan should specify the training the company response team should receive. The safety of the outside emergency responders is the responsibility of the responders' organization and the incident commander. The plan should specify drills, training exercises, or simulation with the local community emergency response planners and responder organizations.

5.4. PRE-STARTUP SAFETY REVIEW

One important function of operations and maintenance personnel is participation in safety reviews. A detailed hazard analysis should be performed near the end of the design phase. Safety recommendations from the previous hazard analyses should be reviewed to find out how the recommendations were resolved. During construction, an equipment arrangement or a process change may have resulted in change orders. The changes may require an additional hazard analysis. The management of change procedures applied to the design changes should have triggered any necessary hazards reviews. During the writing of startup, operating, and shutdown procedures, operating personnel may discover areas that need an additional hazard analysis. Since operations and maintenance personnel will have become familiar with the plant during the pre-startup and commissioning period, hazards analysis techniques[12,13] should be selected that require their participation, as discussed in Chapter 3.

A pre-startup safety review is usually conducted before the startup manager authorizes the initiation of any startup activities. Also, government regulations in the United States require a pre-startup safety review of all new and modified facilities before the unit is started or goes into production.[3] Most companies conduct a pre-startup safety review to verify that appropriate safety precautions have been taken before a plant goes on-stream. Typically, the review is performed by a team composed of production, research and development, engineering, and safety department personnel. It is preferable that the team members have not been involved in the design, construction, pre-startup, and commissioning of the new facility. This approach results in a critical and unbiased review of all previous safety concerns for the construction and anticipated operation of the facility. The pre-startup safety review is conducted after the operating manual, including the operating procedures, is complete and before the completion of commissioning.

---

3 Specific guidelines for pre-startup safety review for new and modified facilities are given in U.S. Department of Labor regulation [29 CFR 1910.119 (i)][4]. A summary of the requirements is given in Appendix A.

The general starting document for the safety review team is the operating manual. The team relies on the documentation of potential hazards associated with the process. The hazards information will cover such items as chemical toxicity, reactivity, flammability, and special problems with materials. The team will be interested in general considerations, such as the design criteria for sizing safety relief valves for exothermic reactions, fire exposure, and thermal expansion. The team typically reviews the alarms and interlocks on critical process variables and all emergency safety systems.

The procedure for conducting the pre-startup safety reviews depends on the safety team members. The team reviews the potential hazards and how the plant personnel and adjacent communities are protected by the process design, equipment layout and spacing, active and passive safety systems, safety procedures, operating procedures, and personnel training.

The team interviews the operations and maintenance personnel to determine their readiness to safely operate the plant. After reviewing and discussing the process drawings to determine the critical areas, inspections of process equipment and plant facilities are conducted with a plant representative. Following the inspection and interviews, the deficiencies are summarized and ranked in importance in a written report. The startup team then reviews the recommendations and a program and schedule are prepared to resolve each of the agreed-on deficiencies.

5.5. COMMISSIONING

The objective of commissioning equipment is to find out if it will perform as intended before process materials are introduced. The commissioning of utilities and equipment is the final step of the pre-startup plan. All inspections, testing, and commissioning should be performed according to established safety procedures. Before commissioning, all vessels, piping, and equipment should be inspected to verify that construction is complete and all items have been tagged or labeled.

To ensure a complete inspection, detailed inspection checklists should be prepared for each item on the master list of all vessels, piping, and equipment. The specific inspection checklists should cover safety considerations such as the accessibility of valves, the generation of static electricity by free falling streams, sampling provisions, and the protection of sensors and sight glasses. Auxiliary items such as foundations, anchor bolts, pipe supports and hangers, insulation, and protective coatings should be covered in the detailed checklists. During the inspections, necessary baseline corrosion data of metal thickness should be recorded. Following the inspections, vessels, piping, and equipment should be pressure tested, cleaned, flushed, dried, and purged. Depending on the process, oxidizing agents, acids, caustics, metal contamination, air, or water can be the source of problems during startup. The importance of proper cleaning, flushing, and purging is illustrated in the following example. During the water batching of an amine absorber column, the flow from the column sump could not be obtained. The column had to be shut down and opened for inspection. When the column sump was entered, a hard hat was found to be plugging the bottom outlet.

5.5.1. Commissioning Utilities

Electric power, water treating, cooling water, service air, steam, condensate, instrument air, inert gas, fire water, underground drains, and fuel oil or gas should be commissioned. Table 5-2 is a general checklist for commissioning utilities. A detailed checklist should be prepared using vendor instructions for commissioning each of the plant utilities. In addition to cleaning, flushing, and purging the lines and headers, each utility should be, if possible, tested for its required delivery capacity.

5.5.2. Commissioning Equipment

Fired heaters; pumps; compressors; vacuum equipment; electric motor, steam turbine, gas-engine and gas-turbine drivers; and their instruments and controls should be commissioned before startup. Commissioning includes such tasks as drying out the refractory or other linings, checking the piping strain on equipment, checking the rotation of electrical motors, and checking the lubrication systems. Equipment critical to safety may need a more extensive commissioning than other equipment. For example, for most non–safety-critical centrifugal pumps, the checklist in Figure 5-8 will suffice. For a safety-critical pump, the checklist should be augmented to include such checks as the correct materials for the pump housing and impeller, correct impeller and wear rings, tandem seals with internal alarms, steam snuffing or fire water spray for seals, and lubrication.

In addition to commissioning individual pieces of equipment, as many process stage and operations as possible should be simulated with water, air, or other inert fluids. If appropriate, water should be pumped through the process. Compressors and blowers should be operated on air or inert gas. Because the process will not have been designed to operate with air or water, the startup team should be careful to avoid damaging the equipment. They should be familiar with the design limitations of the process, including maximum operating temperatures and pressures, the settings of rupture disks and relief valves, and pump limitations. Closed loop dynamic testing with safe fluids permits the flow testing of equipment, indicates how the controls respond, and acquaints the operators with the equipment before process fluids are introduced. Dynamic testing can also be used to collect baseline data such as air flow rates, pressures, and tank flow verification. At the end of the simulation the equipment should be dried and purged.

FIELD INSPECTION CHECKLIST

For_____ Job No._____

Location _____Date_____

No.	Centrifugal Pump	Yes	No	N/A	Remarks
1.	Does installed pump meet specifications?				
2.	Has pump been lubricated?				
3.	Does pump have a drip pan?				
4.	Is correct driver installed?				
5.	Does driver conform to specifications?				
6.	Has driver been lubricated?				
7.	Does coupling conform to specifications?				
8.	Has coupling been lubricated?				
9.	Do installed seals meet specifications?				
10.	Does driver, coupling and pump hand rotate?				
11.	Are pump coupling and shafts adequately guarded?				
12.	If electric driver, are installed breakers and heaters correct?				
13.	Correct rotation?				
14.	Are drain plugs installed?				
15.	Are valves in place for easy operation and service?				
16.	Are there suction and discharge pressure gauges with isolation valves?				
17.	Are flanges and connections tight?				
18.	Are valve packing glands pulled down tight?				
	Water Batching				
1.	Does pump flow have to be restructured to avoid motor overload during water batching?				
2.	Is any part of pump experiencing undue vibration?				
3.	After water batching, has pump been drained, dried and purged?				

FIGURE 5-8. Example field inspection checklist for a centrifugal pump.

5.5.3. Instruments, Computer, and Control

The numerous sensors, analyzers, interlocks, and alarms should be commissioned before startup. Instruments and controls designed as critical to safety should be commissioned before startup. Some of the checks that should be performed are:

- Instrument placement according to drawings
- Specified instrument installed
- Installation according to manufacturer's instructions
- Instrument suitable for electrical classification area
- Calibration
- Continuity check
- Loop check
- Alarms check
- Interlocks check
- Analyzers check

Once the location, instrument, installation, and tagging have been verified to be correct, the instruments should be operated. Scheduling of the inspection and testing should be made from the master list of instruments. For each type of instrument, a checklist should be prepared for testing and documentation purposes. Sensors should be tested by comparing at least two measurements over the control range at the sensor with the readings in the control room. Gas detectors should be calibrated with standard gas mixtures.

The alarm settings and the response of alarms and interlocks to sensors should be verified. The controllers can be operated locally in the plant or remotely by a control room computer via a data highway. Each control loop should be checked. Figure 5-9 provides an example checklist for an instrument control loop[14]. The checking may require bypassing the input output of the computer to verify the logic programmed in the computer. In case a backup computer is installed, its operation should be verified by simulating the failure of the primary computer. Special procedures are needed to verify the operational status of the control system software[15].

5.6. FINAL PREPARATIONS FOR STARTUP

Safety reviews are performed, starting with design and continuing throughout the life of a plant. After the utilities and equipment have been commissioned and the instruments and control systems have been verified, several task and checks should be performed before startup. An example of a pre-startup safety review checklist is shown in Table 5-7.

The startup of a new chemical plant represents a large effort. Detailed pre-startup planning and thorough commissioning of the piping, vessels, and equipment by operations and maintenance personnel is the best way to ensure that the startup will proceed smoothly and safely.

INSTRUMENT LOOP CHECK SHEET

Client: Plant:

Client's Project No.:   Project No.:

Loop No. _____   Service_____

Line or equipment No._____ Pipe I.D._____

Mechanical Checks/Electrical Checks

Measuring element:	Installation correct	☐	Location correct	☐
	Isolating valves correct	☐	Materials correct	☐
	Tapping(s) position correct	☐	Orifice diameter	
Impulse connections:	Correct to hook-up	☐	Materials correct	☐
	Pressure tested	☐	Test pressure	
	Steam/elect. traced	☐	Lagged	☐
Field instrument(s):	Installation correct	☐	Air supply correct	☐
	Weather protected	☐	Power supply correct	☐
Panel instrument(s):	Installation correct	☐	Air supply correct	☐
	Scale/chart correct	☐	Power supply correct	☐
Control valves:	Installation & Location correct	☐	Size & type correct	☐
	Stroke tested	☐	Positioner checked	☐
	Limit switch(es) set	☐	I./P. transducer checked	☐
Air supplies:	Conns. correct to dwgs.	☐	Blown clear & leak tested	☐
Transmission-pneumatic:	Lines inspected, blown clear & leak tested			☐
Transmission-electric:	Insulation checked-core to core	☐	Core to earth	☐
	Continuity checked	☐	Loop impedance checked	☐
	Earth bonding checked	☐	Zener barriers correct	☐
Temperature loops:	T/C or R/B checked	☐	Cable to specification	☐
	Continuity checked	☐	Loop impedance checked	☐
General:	Supports correct	☐	Tagging correct	☐

Checked by: Date   Witnessed by: Date

Loop test:

MEASUREMENT	Transmitter Input	Transmitter Output	Local Inst. Reading	Panel Inst. Reading	

CONTROL	Controller Output	Transducer Output	Valve Pos'nr Output	Control Valve Position	

Remarks:

Checked by:	Date:	Witnessed by:	Date:
Accepted by:	for		Date:
			Instrument Loop No.

FIGURE 5-9. Instrument loop check sheet.

TABLE 5-7
Example of a Pre-startup Safety Review Checklist

- Have all hazard review recommendations been addressed?
- Has the safety critical equipment list been prepared?
- Has all the critical safety equipment been properly installed and is it functional?
- Have the operating manuals been completed and approved?
- Have standard operating procedures been written, including startup and shutdown procedures?
- Have the operators and maintenance personnel received orientation on the process and health hazards of the plant?
- Have the operators been sufficiently trained in the operating procedures for the plant?
- Are field change authorization procedures in place?
- Has all computer control logic been thoroughly tested?
- Has an emergency response organization been assembled and have responsibilities been assigned?
- Has a written emergency response plan been prepared and have practice drills been run?

5.7. REFERENCES

1. Finlayson, K. and Gans, M. "Planning the Successful Startup." *Chemical Engineering Progress,* Vol. 63, No. 12, December, 1967.
2. Synthetic Organic Chemical Manufacturer's Association. "Chemical Process Operator Certification Training Information Packet." Washington, DC. 1992.
3. Matley, Jay. "Keys to Successful Plant Startup." *Chemical Engineering,* September 8, 1969.
4. *Process Safety Management of Highly Hazardous Chemical; Explosives and Blasting Agents; Final Rule,* 29 CFR Part 1910. Department of Labor, Occupational Safety and Health Administration, Washington, DC, February 24, 1992, *Federal Register,* Vol. 57, No. 36, pp. 6356-6417.
5. Stainthorp, F.P. and West, B. "Computer Controlled Plant Startup." *The Chemical Engineer,* September, 1974.
6. Guide for the Development of State and Local Emergency Operation Plans (CPG 1-8). Federal Emergency Management Agency, October 1985.
7. U.S. Department of Transportation. Hazardous Materials Emergency Response Guidebook (DOT-P-5800.5). 1990.
8. Chemical Manufacturer's Association. Community Awareness and Emergency Response Program Handbook.1986.
9. 29 CFR Part 1910.120—Occupational Safety and Health Standards. Section 1910.120, Hazardous Waste Operations and Emergency Response, 54 FR 9294, March 6, 1989.
10. 29 CFR Part 1910.138—Occupational Safety and Health Standards. Section 1910.138 (a), Emergency Response Plan.
11. 29 CFR 1910.165,—Occupational Safety and Health Standards. Section 1910.165, Employee Alarm Systems.
12. Center for Chemical Process Safety (CCPS). *Guidelines for Technical Management of Chemical Process Safety.* American Institute of Chemical Engineers, New York, 1989.

13. Center for Chemical Process Safety (CCPS). *Plant Guidelines for Technical Management of Chemical Process Safety.* American Institute of Chemical Engineers, New York, 1992.
14. BS6739, "Code of Practice for Instrumentation in Process Control Systems: Installation Design and Practice." British Standards Institution, 1986.
15. Horsley, D.M.C and Parkinson, J.S. (Editors). *Process Plant Commissioning—User Guide.* Institution of Chemical Engineers, London, 1990.

6

STARTUP

Startup is the critical transition period when the plant is taken from a shut-down state through a predetermined sequence of steps to an operating state. Although startups are infrequent and their durations are short compared with the life cycle of a plant, process safety incidents occur five times as often as they do during normal operations.[1] This is because the startup period is intense and the entire startup team is under stress to bring the plant into operation. Because of the infrequency and intensity of startup, operation and maintenance personnel should be alert to the following keys to safe startups:

- Knowing and understanding the startup plan
- Training and drilling on startup procedures
- Avoiding rushing or taking unnecessary risks during startup
- Mitigating the effects of long hours, nonroutine work, and stress associated with startup
- Observing and documenting any changes or deviations from the startup plan or procedures.

The types of startup are:

- *Initial Startup* Initial startup is the sequence of steps taken to bring a newly constructed plant and equipment to operating condition for the first time. Initial startup is sometimes called commissioning. To avoid confusion, we will use commissioning here to refer to the preparation of equipment for operation after construction or replacement is complete.
- *Restart* A restart is a startup after a sudden trip-out or emergency shutdown. A restart can occur for example, after part of a plants piping or equipment has been bypassed for maintenance or repair while the rest of the process is placed in a holding status.
- *Startup after Turnaround* A startup after turnaround occurs after an operating unit has been shut down for required inspections, preventive maintenance, equipment repairs or process modifications.

- *Startup after Extended Outage* A startup after an extended outage is the startup after facility has been idle and unattended.

6.1. ROLES AND RESPONSIBILITIES

The operations department has the lead role in startups. Initial startups and startups after an extended outage are usually performed with the help of a team of startup specialists. The maintenance department's role in startups is to support the operations department by installing and removing temporary piping, making emergency repairs, monitoring, and giving general assistance. The operations and maintenance departments should ensure that any deviations from design or changes made during startup are documented and that appropriate changes are made in the process safety information (e.g., drawings), the operating procedures, and maintenance schedules and procedures.

If an engineering and construction company developed the process, the job may be a turnkey project in which the contractor is responsible for the initial startup of the plant and/or for achieving design production rates. For turnkey projects the operations and maintenance department's roles are usually to observe and support the contractor's startup team. When the process development and design are done by the operating company, the company startup team is responsible for the initial startup. For the initial startup of non-turnkey projects, the operations and maintenance personnel have their greatest responsibility. Before initial startup, whether it is a turnkey project or not, operations personnel will have been trained in the procedures for startup, normal plant operations, and shutdown. The startup team will be in charge of the startup and will direct the operations, maintenance, and other support groups assigned to the startup team until normal operations are achieved. The normal operations and maintenance groups may be augmented by specialists during the startup. Minute-to-minute operations and maintenance will be the responsibility of the startup team. Restarts and startups after turnaround are the responsibility of the operations department.

The maintenance group is essential to a safe and successful startup. The staffing of the maintenance group during startup will affect the speed and safety of necessary repairs and modifications. Quick repairs can often prevent the shutdown of the plant and ensure a smooth and safe startup. The number of maintenance personnel needed will depend on the size of the plant and access to outside personnel. It is best to have backup staff available during initial startups, startups after turnaround, and startups after an extended plant outage. The maintenance department's role for all types of startups is mainly supportive.

6.2. INITIAL STARTUP

To execute a startup safely and efficiently, the startup team should consist of a well- organized, trained staff of experienced people; each detail should be planned and scheduled; and direct communication should exist among the various disciplines and groups. The typical startup team is usually composed of a startup manager, the operations group, the maintenance group, and technical specialists, including a laboratory group. Figure 6-1 shows an example startup team for a process unit in a major chemical plant.[2]

Figure 6-1 shows the members of the startup team who typically form the startup committee. The operations organization shown in Figure 6-1 is the same as for normal operations, except the shift engineers. Although Figure 6-1

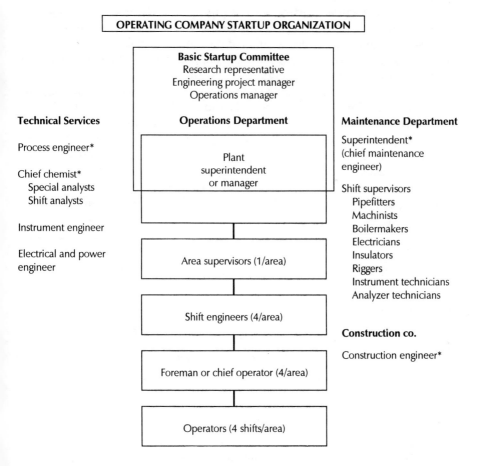

FIGURE 6-1. Example startup organization. Positions marked with an asterisk are included in the expanded startup committee

shows the shift engineers with line authority, in many companies shift engineers serve as consultants. When the shift engineers have staff responsibilities, they can devote more time to technical problems. Examples of maintenance requirements during initial startup are shown in Table 6-1.

Planning and scheduling are the keys to a safe and efficient startup. When planning the startup, the operations and maintenance departments should ensure that each step of the startup is planned and that operating instructions are prepared for each step. Figure 6-2 shows a labor chart that shows the startup steps, the time to start the step, and the projected duration of each step for an ammonia plant. Figures 6-3 (a), (b), and (c) show another method of scheduling the commissioning, pre-startup, and startup activities for a crude unit. In Figures 6-3 (a), (b), and (c) the times to complete the tasks are shown in boxes describing the tasks.

During the initial startup, all the piping, vessels, equipment, utilities, instruments, valves, controls, alarms and interlocks operate simultaneously for the first time with process fluids at design pressures, temperatures, and flow rates. Even if the design, construction, pre-startup, and commissioning phases are completed conscientiously and thoroughly, problems can occur during startup, adding to the stress and pressure on the startup team. Operations and maintenance supervisors should be alert to the tendency to take undue risks at these times. Causes of delays can be grouped into four categories: design inadequacies, construction shortcomings, equipment deficiencies, and operating errors. Table 6-2 on page 124 shows the results of two studies giving the distribution of startup delays. The table shows that equipment problems account for most of the delays, which can be decreased by detailed planning by the maintenance department for spare parts and quick repairs.

TABLE 6-1
Examples of Maintenance Requirements during Initial Startup

Personnel	Number	Responsibilities
Machine fitters/mechanics[a]	1 full time up to five	Operating equipment such as pumps, agitators, centrifuges, and compressors
Pipe fitters and welders	2 full time up to ten	Making repairs and changes to piping and vessels
Electricians	on call	
Insulation installers		Installing thermal insulation for personnel protection
Instrument technicians	1 for each shift	Repairing and adjusting process analyzers and instruments
[a] Must have access to well-equipped machine shop.		

Activities	No. of days																
	1	2	3	4	5	6	7	8	9	10	11	12	13	14	15	16	17
Start feed gas to primary reformer and increase temperatures	■																
Light off secondary reformer and activate high-temperature shift converter	■																
Start gas flow through absorber and line-out system		■															
Activate methanator		■															
Activate low-temperature shift converter		████															
Align synthesis gas and refrigerant compressors			██														
Start refrigerant and synthesis gas compressors, heat converter shell and test for leaks				████													
Heat up and reduce catalyst			██														
Activate ammonia converter and send first product to storage						████											
Increase feed rates to 100% and production to 1,000 tons/day								██████████████									

FIGURE 6-2. Startup operation schedule for an ammonia plant.

117

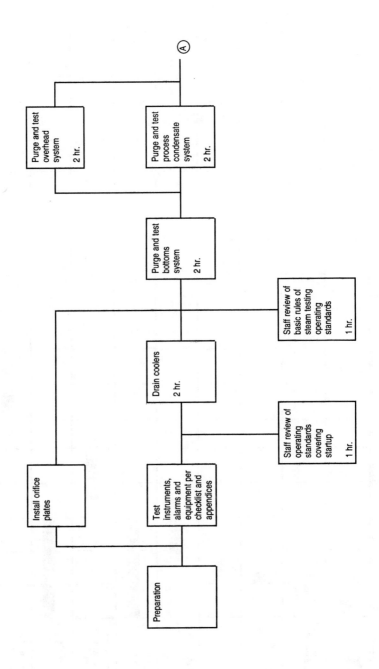

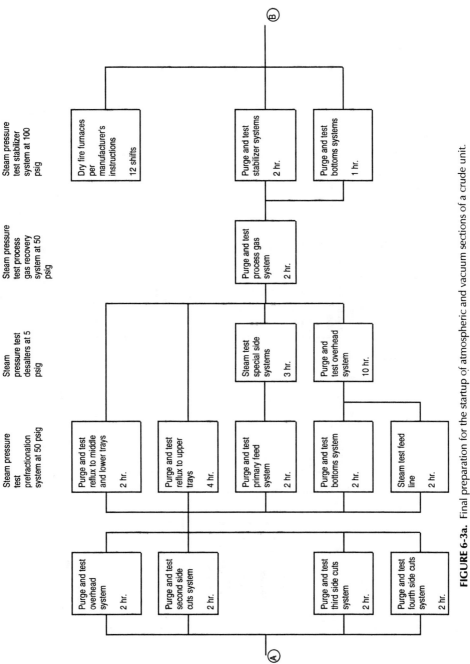

FIGURE 6-3a. Final preparation for the startup of atmospheric and vacuum sections of a crude unit.

119

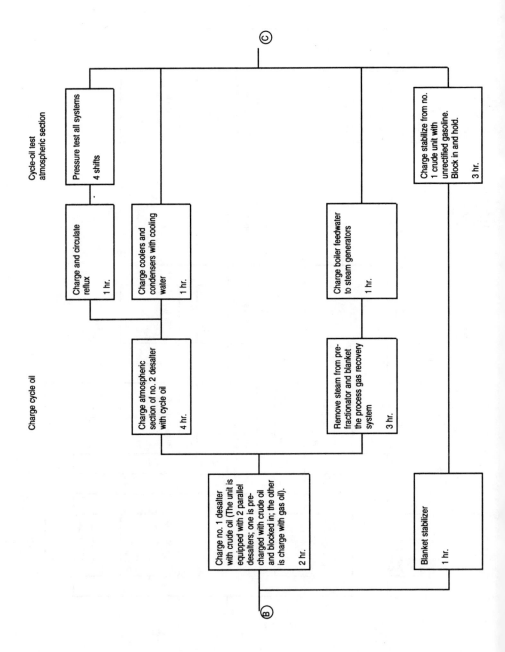

Charge cycle oil

Cycle-oil test
atmospheric section

Pressure test all systems

4 shifts

Charge and circulate reflux

1 hr.

Charge coolers and condensers with cooling water

1 hr.

Charge atmospheric section of no. 2 desalter with cycle oil

4 hr.

Charge boiler feedwater to steam generators

1 hr.

Remove steam from pre-fractionator and blanket the process gas recovery system

3 hr.

Charge stabilize from no. 1 crude unit with unrectified gasoline. Block in and hold.

3 hr.

Charge no. 1 desalter with crude oil (The unit is equipped with 2 parallel desalters; one is pre-charged with crude oil and blocked in; the other is charge with gas oil).

2 hr.

Blanket stabilizer

1 hr.

Ⓒ

Ⓑ

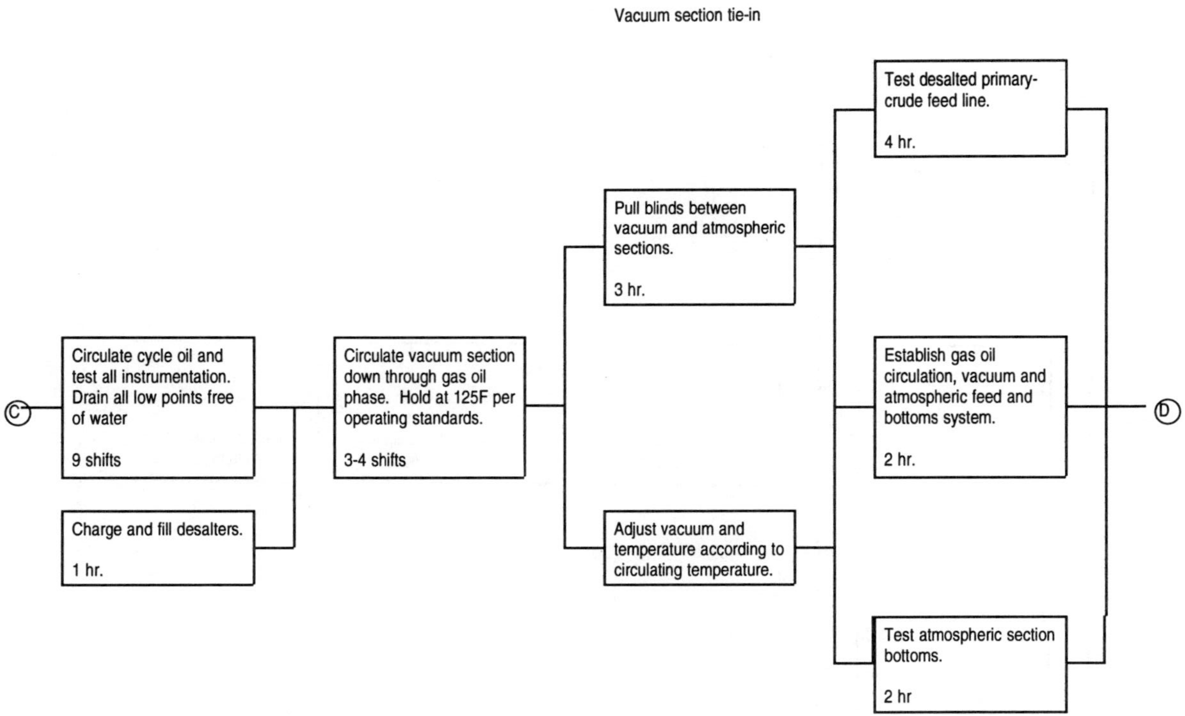

FIGURE 6-3b. Final preparation for the startup of atmospheric and vacuum sections of a crude unit.

Vacuum section tie-in

Test desalted primary-crude feed line.

4 hr.

Establish gas oil circulation, vacuum and atmospheric feed and bottoms system.

2 hr.

Test atmospheric section bottoms.

2 hr

Pull blinds between vacuum and atmospheric sections.

3 hr.

Adjust vacuum and temperature according to circulating temperature.

Circulate vacuum section down through gas oil phase. Hold at 125F per operating standards.

3-4 shifts

Circulate cycle oil and test all instrumentation. Drain all low points free of water

9 shifts

Charge and fill desalters.

1 hr.

121

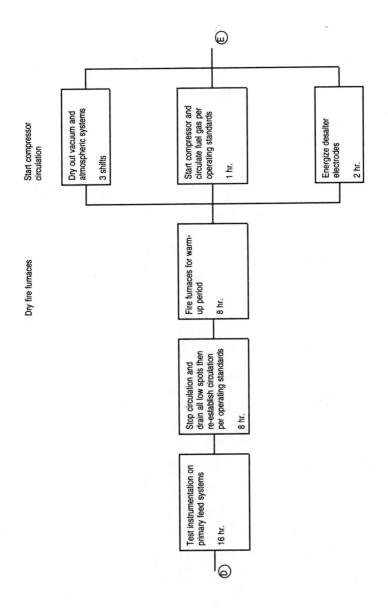

Dry fire furnaces

Start compressor circulation

Test instrumentation on primary feed systems
16 hr.

Stop circulation and drain all low spots then re-establish circulation per operating standards
8 hr.

Fire furnaces for warm-up period
8 hr.

Dry out vacuum and atmospheric systems
3 shifts

Start compressor and circulate fuel gas per operating standards
1 hr.

Energize desalter electrodes
2 hr.

D

E

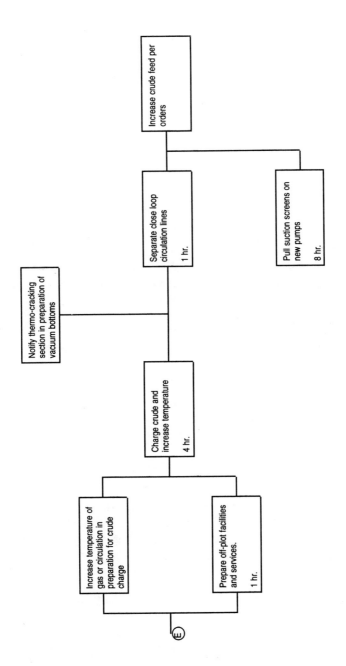

FIGURE 6-3c. Final preparation for the startup of atmospheric and vacuum sections of a crude unit.

TABLE 6-2 Causes of Startup Delays		
	Percent of delays	
Causes	Holroyd [3]	Finneran et al. [4]
Design Inadequacies	10	—
Startup Operating Conditions	—	15
Construction Shortcomings	16	14
Equipment Deficiencies	61	56
Operating Errors	13	15
	100	**100**

6.2.1. Final Preparation

Before process chemicals and materials are introduced into vessels and piping, several checks should be performed. Any deficiencies found during inspection, testing, and commissioning should be corrected. All changes made during this period should be documented and appropriate changes should be made to the process safety information and the operating and maintenance procedures. Categories of hardware that should be checked are:

- Piping, vessels and equipment
- Utilities
- Instruments, computer, and controls
- Pressure relief and flare system
- Alarms and safety interlock systems
- Fire, gas, and smoke detection
- Fire water and chemical fire suppression
- Diking, curbing, and drainage

The startup team should verify that all design safety and pre-startup safety review recommendations have been satisfactorily addressed. The training of the operators, maintenance personnel, and laboratory technicians should be verified. All startup personnel must know the emergency response plan and what action to take in an emergency. The spare parts inventory should be reviewed from the perspective of the maintenance department being able to make repairs quickly. Safety equipment should be inventoried and locations checked for accessibility.

Other areas that should be checked are the written procedures, available personnel, and process information. Procedures should be in place for:

- Safe work practices (confined space entry, hot work permits, lock-out/tagout, etc.)

- Startups
- Normal operations
- Shutdown (normal and emergency)
- Management of change
- Repair
- Maintenance
- Laboratory Work
- Process evaluation

Also, check whether there are enough operators; maintenance personnel; laboratory technicians; instrument, computer, and control engineers; electrical and power engineers; research chemists; process engineers; development engineers; design engineers; construction engineers; and shift engineers. The process safety information should be accessible, and the users of the information should know where it is and how to find it before and during startup. The operating manual, all process hazard information, and as-built drawings should have been updated to reflect any changes or deficiencies discovered during pre-startup and commissioning.

6.2.2. Introduction of Process Chemicals and Materials

Initial startup begins when process chemicals and materials are introduced into the piping, vessels, and equipment. During this period, the operations department has the lead role in directing the charging of the initial inventory; the maintenance department supports operations by providing temporary piping, making emergency repairs, monitoring, and giving general assistance. Since the introduction of process chemicals is a major transition point, it is important that operations and maintenance personnel understand the physical properties, reactivity, polymerization (decomposition), corrosiveness, flammability, and toxicity of the process chemicals. Their process knowledge, combined with the available detailed procedures is the main element of safely handling process chemicals. Detailed checklists of the alignment of pumps, valves and vessels should also be part of the documented procedures, which should clearly specify checking the condition of the piping, vessels and equipment before introducing process chemicals.

Residual materials, such as water and air remaining from the commissioning phase, may not be compatible with the process fluids in the section of the plant being charged or with downstream parts of the process. Procedures should be developed to remove residual materials and avoid possible contamination, side reactions, and corrosion that could result in unsafe conditions or nonspecification product.

The task of charging the initial inventories is usually integrated with the startup procedures. The startup operating procedures for introducing initial process materials should be followed precisely. Typical pieces of equipment that startup personnel have extensive experience operating and maintaining may require atypical procedures during the initial charging. For example, a distillation column is usually initially charged by introducing feed into the column sump and heating the reboiler section with the reboiler on total recycle. After the reboiler has reached the operating temperature, feed is slowly introduced while the temperature is maintained in the lower section of the column. The column continues to be charged until a level can be maintained in the reflux drum when the unit is operated under total reflux. The column is usually operated on total reflux until other parts of the plant are brought on line to provide feed to the column. Downstream units are often charged with fluids that have compositions near those anticipated in the process.

Catalysts for fixed and fluidized beds should be in the reactors before the reactants are introduced. Pelletized or granular catalysts are usually introduced in both fixed and fluidized bed reactors as the final steps in the equipment commissioning process. For fluidized beds, the catalyst charge is usually less than the full charge. As the process flow rates are increased, additional catalyst is added to reach the optimum bed depth. For a reactor using liquid catalysts, such as hydrofluoric acid for alkylation, the feed tank is charged and the catalyst is added as the reactants are introduced into the reactor. Initiators are usually introduced with the reactants, and since large quantities and continued makeup are not required, the initial charge may be made from temporary storage. An example of an initiator is sulfuric acid, which causes a cleavage reaction to cumene hydroperoxide, forming acetone and phenol. Once the reaction starts, it continues without additional sulfuric acid. The most common characteristics of liquid catalysts, initiators, and caustics are their reactivity and corrosiveness, which the operations and maintenance personnel introducing these materials into the process should consistently remember.

During initial startup, operations and maintenance staff must ensure safe conditions for the transport and transfer of process materials from trucks, rail cars, barges, ships, or pipelines. These materials can include lubricants, process fluids, reactants, catalysts, caustics, initiators, and water treatment chemicals. The process fluids and reactants will usually be transferred to a feed tank and then to the process vessels. Detailed checklists and procedures will help ensure that the fluids are isolated in the correct feed tanks and process vessels. Where transfers are made with temporary piping, additional precautions should be taken for handling the process materials.

6.2.3. Process and Process Equipment Monitoring

During startup, deviations from normal operating conditions can result in hazardous incidents and accidents. To start the process and equipment, however, it may be necessary to operate outside normal operating conditions;

therefore, close monitoring of the process and equipment is essential. For example, a low-liquid-level alarm may be temporarily disabled until the liquid fills the previously empty tank to within the normal operating range. Startup difficulties are more often caused by operators forgetting design and construction details than by fundamental design errors. Operations and maintenance personnel monitor such details that contribute to the general safety and efficiency of the startup. Relying on their experience with process plants, operating and maintenance personnel can perform a major role in troubleshooting during startup. Because of their "hands-on" contact with the process and equipment, operations and maintenance personnel will often be the first to know of problems involving:

- Leaks
- Fluid flow
- Vibration
- Equipment mounting
- Phase separation
- Controls
- Equipment suitability

Many hazards present during equipment commissioning must also be guarded against during startup. For example, most equipment and piping will have been pressure tested, but when piping and vessels are pressurized during startup, unnecessary personnel should be cleared from the area. Leaks that do not occur during pre-startup testing may appear because of vibration or temperature changes. Leaks of high-pressure, toxic, or flammable substances can put personnel, equipment, or the process at risk. Figure 6-4 describes an example of a small leak that turned into a major problem during the startup of a refinery after a major process revamp.

Leaks of incompatible material into the process stream, although usually not as obvious as external leaks, can create havoc with a startup. For example, if air, water, oil or grease is slowly leaking into the process stream, towers will flood and foam, heat exchangers will foul, pumps will cavitate, and routine chemical analyses will be upset. Other fluid flow problems that may occur are the short circuiting of fluid vessels because of inadequate baffling, the slugging of pumps caused by a lack of gas–liquid separation or gas venting of the inlet streams, and the cavitation of pumps caused by an insufficient net positive suction head.

Changes in the vibrations of major rotation equipment are usually monitored via specialized instruments; however, operations staff should routinely monitor and double check all equipment. The vibrations of piping and surrounding structures can usually be observed by operating and maintenance personnel. Mounting stirrers separate from the vessels can cause alignment, sealing and vibration problems.

Factual description:
During startup, after a major process revamping in a refinery, as the reboiler section of the debutanizer was being brought up to temperature, a leak developed through the gasket of the reboiler head. The leak produced a small, hot oil mist cloud. While bolts were being tightened on the head of the reboiler, the feed to the column was diverted to an empty atmospheric storage tank at the edge of the tank farm. While the feed was being diverted to the storage tank, some feed "flashed," causing mostly butane gas to overflow the tank and the diked area. The heavier-than-air vapor cloud surrounded a nearby house and spread across a road, where it was ignited by a car. The fire traveled back to the storage tank, resulting in a fire that collapsed the floating roof and caused extensive heat damage to the tank. This example shows how a minor leak can lead to a major incident when the response to an emergency has not been planned and poor decisions about changes are made under the stress of startup.

Contributing causes	Process safety concept
1. The emergency response plan should have included proper instructions about adequate response to a leak to avoid a small leak leading to a catastrophic situation.	Emergency response procedures
2. The process hazards analysis should have identified the potential of the butane flashing during its transfer to the storage tank.	Process hazards analysis
3. The process hazards analysis findings should have been used to develop adequate operating procedures and train personnel.	Operating procedures, training

FIGURE 6-4. Refinery revamping startup incident.

Because of the uncertainty associated with the scale-up of pilot plant data for separation equipment, operations and maintenance personnel should be alert to potential problems with gas-liquid–solid separations, such as settling, filtration, scrubbing, and demisting. In some water-hydrocarbon settling separations, biological growth at the interface can cause problems with sensors such as electrical conductivity sensors.

The performance of controls is usually partially tested during water batching during commissioning. In the presence of process fluids and normal reaction conditions, the controls may not function as planned. Examples of difficult controls to maintain are pH, slow response between action and detection, and control loops based on nonstandard variables such as electrical and thermal conductivity. Operations and maintenance personnel should be prepared to implement other ways of operating. Figure 6-5 is an example of the importance of controls, of adhering to operating procedures, and of maintaining equipment integrity.

The startup may reveal pieces of equipment that are not suitable for process duty and that require a high rate of maintenance. Before assuming that the equipment is unsuitable or defective, abnormal conditions during startup should be considered. The running-in period for some equipment may be different from familiar normal operation. For example, high concentrations of metal particles may be present in lubricating oils in new equipment. The

Factual description:
In a batch process to make a herbicide intermediate, cyanide was used as one of the reactants. The process was set up on a programmable logic controller that stepped through the process sequence. The controller called for the addition of several reactants and then started the addition of cyanide. The hydrogen cyanide flow totalizer was improperly calibrated and the amount of cyanide was significantly more than was required for the batch. The batch overheated and blew out the rupture disk, releasing cyanide. It was later determined that the flow totalizer had never been used in cyanide service and was incapable of correctly measuring cyanide flow. Operations continued to operate the process by adding cyanide to the batch and observing the color of the reaction mix by shining a flashlight through an observation port of the reactor. The cyanide addition was controlled in this manner for over a year until an acceptable instrument could be installed.

Contributing causes	Process safety concept
1. The mechanical integrity program should have identified the flow totalizer as a critical piece of equipment which would have required periodic inspections and maintenance of the flow totalizer.	Mechanical integrity
2. Unmanaged change in operating practices, e.g., use of the flashlight through the observation port to observe the color of the reaction mix.	Management of change Audit program
3. The audit program for the facility should have identified the unmanaged change or the deficiency in the mechanical integrity program.	

FIGURE 6-5. Performance of controls.

vendor or manufacturer should be able to provide information about the normal break-in condition of lubricating oil in the equipment.

The inexperience of process operators with the equipment, design, or process being started may result in damage to the equipment and increased risk to the safety of personnel. The improper operation of equipment during startup can shorten the life of the equipment and decrease the integrity of the process. For example, operating equipment at higher-than-normal temperatures can cause an increase in the creep rate. It is possible to use up a large portion of the equipment creep life during startup. Damaging equipment by improperly operating it is more likely to occur during startup. Allowing a pump to run dry or to pump against a closed valve would be unlikely during normal operations but can easily occur during startup.

Some seemingly unimportant differences in startup and normal operations may cause problems if they are not monitored. For example, using nitrogen for purging may become a problem during startup. The greater-than-normal volume being used in a particular area could become an asphyxiation hazard. Utility use during startup may be excessive, increasing risks in other areas.

The startup period is characterized by peak physical and psychological stress. The increased physical activity is caused by the need to make careful observations, overcome equipment failures, act in place of nonfunctioning automatic controls, and work extended hours. Psychological stress is caused by the unfamiliar working conditions, the need to make quick decisions, time constraints, and the need for heightened awareness. Everyone involved should be aware that hazardous conditions are more likely to occur during startup; this awareness not only promotes caution but causes stress. The first- and second-line supervisors should make every effort to enhance safety and decrease the stress on personnel during the startup period.

The stress and long hours associated with a difficult startup may cause personnel to accept undue risks and fail to keep complete records. To ensure the safety of the startup and the continued operation of the plant, it is essential that any abnormalities observed during startup be communicated and well documented. The operations and maintenance departments should ensure that any changes or repairs made during startup are carefully documented.

6.2.4. Baseline Data

Data should be recorded on a schedule during the startup. It is important to record as much relevant data as possible to check the design calculations and keep a log of construction faults, equipment deficiencies, and shortcomings in the startup procedures . Sometimes the plant cannot be operated at design conditions or produces better yields and throughput under other conditions experienced during the startup. A record of these operating conditions will increase the staff's understanding of the process and help them change the alarm and interlock settings in the startup and operating procedures. Should operating problems develop, data should be recorded as frequently as possible. Although problems during startup are stressful, it is essential to record data and document abnormalities in order to make process equipment, control and operating document changes later. For startups that prove extremely difficult, supervisors should consider adding a person to each shift to record data, allowing the operators to focus on the process. When an emergency shutdown is necessary to reduce the risk to lives and equipment, the data and observations can be analyzed to find the cause of the shutdown. When evaluating process safety, differences in the recorded data from the design-predicted values should be resolved.

When the cause of the problem cannot be understood from the operating data, special tests should be performed to obtain additional data. Operations and maintenance personnel should be prepared to set up complete recordings for temperatures, pressures, and process flow rates in the problem area to obtain additional data for analysis.

6.2.5. Updating Startup Procedures

As the process is being started up, documentation should cover any problems with or failure of: (1) equipment; (2) the safety protective systems, particularly sensors, alarms and interlocks; (3) maintaining critical process parameters within design ranges; (4) controls; and (5) procedures. The startup procedures should be changed as soon as possible, either during a shutdown or after the process has stabilized at design rates and specifications. Changes that require only clarification of the startup and operating procedures may be made by revising the operating manual. All revisions should carry the date of the latest revision.

Data obtained during the startup can be used to understand the process and may result in changes being made to the critical operating parameters. However, before revising any of the critical operating parameter limits, such as temperature, pressure, or fluid flow rates, operations supervisors should initiate a management of change review to evaluate the effect on process safety. Process, equipment, instrumentation and control changes should trigger management of change procedures that will include necessary revisions to the startup, operating, and maintenance procedures.

6.3. RESTART

A restart is a startup after a sudden trip-out or emergency shutdown. Sometimes the part of the process that was the source of the problem is bypassed for maintenance or repair and the rest of the process is placed in an idling status until repairs can be completed and the process restarted. After the idled process has been stabilized according to the operating instructions, operations and maintenance personnel should consider several process safety issues either to prepare for restart or to complete the shutdown by reducing, isolating and decontaminating the inventory. First, operating staff should find the cause of the shutdown. Next, maintenance personnel should estimate the time needed to complete any necessary repairs. On the basis of the time of repair estimate, operations personnel should decide if the process can be safely held in an idling state during the repair period and if the process can be safely restarted.

To safely hold the process in an idling state, supervisors should consider the adequacy of normal operation instrumentation to control the process. For example, a low flow rate may be below the lower detectable limit of a pressure or flow controller. Also, the effects of reactive materials and catalysts should be considered. The retention of monomers in reaction vessels for extended periods could affect the degree of polymerization and the rate of heat removal during the repair. The increased polymerization could result in an elevation of the boiling point and a reduction of the heat transfer coefficient that could cause restart problems and the production of nonspecification product. While repairs are being made, operations also should plan where in the startup sequence to begin and review the deviations in the startup sequence that will be required for the restart.

6.4. STARTUP AFTER TURNAROUND

A startup after turnaround occurs after the plant has been shut down for required inspections, preventive maintenance, equipment repairs, or process modifications. During a startup after a turnaround, operations staff will not have the benefit of the detailed planning and extensive expertise available during the initial startup. The piping, vessels, and equipment disassembled during the turnaround will have to be inspected, tested, and commissioned. Critical safety systems related to the process piping, vessels, and equipment repaired during the turnaround should be recommissioned before startup. Any process equipment idled during the turnaround should be inspected and tested. When bolts are tightened on a flange, they stretch. After heating during the process they lose the ability to stretch. The opposite of this is that they lose their ability to shrink. When the unit is down, the piping cools off, and the tension of the bolts on the flange is reduced. Therefore, when the process is restarted and pressurized before it reaches operating temperature, it is likely that some flanges in the unit will leak. This is especially true in a high-pressure system, in large piping, and in systems containing components that are difficult to contain, such as hydrogen.

Another problem is that the mass of large, high-pressure flanges will heat up more slowly than the rest of the piping, which means that the bolts will heat up even more slowly. This can cause the bolts, during a reheating stage, to become overtorqued and over stretched. When the heat load is evened out, the overstressed bolts may not shrink back to the required torque and can cause the flange to leak. Many other process equipment effects during shutdown and restart can cause safety problems. Problems that may exist for each type of process and equipment should be identified and a checklist should be developed for the time the unit is down. Whenever experience shows that problems exist specific to a unit during shutdown and startup, the checklist should be revised to reflect the experience.

It is important that the commissioning of repaired equipment be as thorough as the commissioning for an initial startup. An example demonstrating the importance of adhering to operating procedures and mechanical integrity programs is given in Figure 6-6.

Process and equipment modifications require management of change procedures. First- and second-line supervisors of the operations and maintenance departments have the lead role initiating and executing management of change procedures during the turnaround. Operations personnel will be

Factual description:
During a turnaround, the instruments and controls were overhauled and checked on a large refrigeration compressor train. The compressor train consisted of a 20,000-horsepower, first-stage compressor and two separate 10,000-horsepower, second-stage compressors. Flow through the first-stage compressor was controlled by a variable speed drive on the steam turbine driver. The second-stage compressors had electric motor drivers and were controlled by flow control valves on the suction of the compressors. Figure 6-7 shows a simplified control loop for the compressor train. On reinstalling the instruments and controls on the second-stage compressors, the valve action was reversed on one of the compressors. During commissioning of the compressor train, the control loop was checked at a mid-range output for the flow and both control valves showed nearly identical valve openings. During the startup the first of the second-stage compressors was brought on line with no problem. As the second compressor, the one with the reversed valve action, was brought on-line and the flow increased through the train, the controls of the first- and second-stage compressors began fighting each other and caused cycling. The undamped cycling caused the first-stage compressor to surge. During the surge, the flow was lost from the first stage and the high pressure on the suction side of one of the second-stage compressors caused the check valve, the backflow preventer, to slam shut. The body of the check valve fractured, causing a large propylene spill. This example illustrates the need to check control loops over the entire operating range during commissioning to prevent startup problems.

Contributing causes	Process safety concept
1. The valve action on one of the compressors was reversed during reinstallation.	Operating procedures, training
2. The pre-startup safety review should have included a check of the control loop over the entire operating range.	Pre-startup safety review

FIGURE 6-6. Turnaround startup incident.

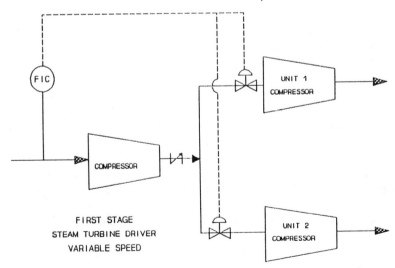

FIGURE 6-7. Simplified control system on a propylene compressor train.

responsible for making necessary updates about any changes in the operating procedures to the operating procedures training, and documentation of the training. Maintenance staff should make any necessary changes to maintenance procedures, schedules, and the inventory of spare parts.

Before a startup after a turnaround, a pre-startup safety review should be performed if the modifications trigger management of change procedures. The pre-startup safety review, similar to the review described in Chapter 5, should be applied to the parts of the plant affected by the turnaround modifications and management of change criteria.

After the turnaround repairs have been completed, the piping, vessels, and equipment have been inspected and commissioned, and final preparations for startup have been completed, the startup sequence should follow the same sequence as the initial startup. One exception may be the introduction of process fluids. Some piping, vessels, and equipment may not have been affected by the maintenance, repairs, and modifications of the turnaround and were isolated, along with their inventories of process fluids. Although the process had been started up and run previously, the introduction of process fluids and the initial process monitoring should be given the same degree of attention as during the initial startup.

6.5. STARTUP AFTER EXTENDED OUTAGE

The startup of a process after it has been idle and left unattended should be treated as an initial startup although some additional safety precautions should be taken by operations and maintenance personnel. All the steps for inspecting, cleaning, testing, and commissioning should be performed on all the piping, vessels, and equipment being returned to service. Figure 6-8 is an example of the failure to perform an inspection and the resulting incident. The example shows that inspection and commissioning steps can help operators avoid incidents after an extended outage.

The deterioration and accumulation of rust in piping, vessels, and equipment should be noted. During flushing or chemical cleaning, unexpected materials trapped in the system may react, causing the formation of hazardous gases. An example is the formation of hydrogen sulfide when hydrochloric acid encounters iron sulfides. Depending on the length of time the process has been out of service and on the history and documentation of any process hazard analyses, a process hazard review may be necessary. In addition, a pre-startup safety review should be performed before startup. The pre-startup safety review will determine if the operating procedures are up to date and if the training of the operators is sufficient and documented. A check of mechanical integrity, maintenance procedures, and maintenance personnel training should be included in the pre-startup safety review.

Factual description:
A BTX unit (benzene, toluene, and xylene) was being recommissioned. To meet specification, ethylbenzene was acid washed in a liquid-liquid column and clay filtered to remove some of the color. Later it was decided that the acid-washed ethylbenzene was causing problems in the styrene manufacturing and the acid wash column was shut down and mothballed. After about eighteen months, a decision was made to go back to the acid wash. The column was checked out and returned to service. No internal inspection was performed. The column would not operate satisfactorily and was shut down, cleaned out, and internally inspected. On inspection it was discovered that all the downcomers for the flow baffles had corroded internally and all the baffling in the column had collapsed. The downcomers, trays, and baffles had failed because of corrosion.

Contributing causes	Process safety concept
1. A process hazards analysis should be performed before the recommissioning. The process hazards analysis should then be used to determine what inspections are needed.	Process hazards analysis
2. The procedures and training program should have required a pre-recommissioning process hazards analysis and internal inspection.	Operating procedures, training

FIGURE 6-8. Extended outage startup.

6.6. RESOURCES

Planning, preparation, and communication are the keys to safe and successful plant startups. The organization and ease of access of information are important to the solution of problems that occur during startup. Besides specific process and company procedures, several types of reference material should be available, such as:

- Process Safety Management[5,6]
- Loss Prevention[7]
- Process Plant Startup[8]
- Chemical Engineering Handbook[9]
- Properties of Hazardous Materials[10]
- Mechanical Engineering Handbook[11]
- Data Books on Gases,[12] Liquids, and Solids.

Part of the planning and preparation of the startup and commissioning phase of a plant should include the organization and creation of a plant-specific information center. The information in the information center may include:

- Design basis
- Relief system and design basis
- Ventilation system design

- Material and energy balances
- Detailed process description
- Process flow diagrams (PFDs)
- Piping and instrument diagrams (P&IDs)
- Electrical drawings and classification areas
- Plot plans
- Process chemical safety data
- Safety procedures
- Management of change procedures
- Startup, operating, and shutdown procedures
- Inspection, tagging, and commissioning procedures
- Detailed equipment specifications
- Utility use
- Vendor's drawings, operating instructions, and correspondence
- Engineering design standards
- Safety systems (interlocks, alarms, detection, and suppression)
- Fire protection system
- Process hazard reviews
- Analytical procedures
- Inspection and commissioning records
- Operator and maintenance training data
- Routine maintenance schedules
- Mechanical integrity inspection and testing program
- Instrument and alarm settings
- Calculation procedures for plant performance
- Training plan
- Startup and commissioning plan
- Emergency response plan.

Part of this information is usually assembled into an operating manual with the following sections:

- Detailed Process Description
- Description of Utilities
- Process Equipment
- Preparation for Startup
- Operating Procedures
- Safety and Emergency Procedures.

The operating manual is then used as the specific guideline for starting and operating the plant.

The initial startup of a plant requires that staff training and the number of the permanent working force of operators, maintenance personnel, and laboratory technicians are adequate. Maintenance personnel include:

- Pipefitters
- Machinists
- Boilermakers
- Electricians
- Insulators
- Riggers
- Instrument technicians
- Analyzer technicians.

Other temporary-task force personnel that may be needed for the startup are:

- Instrument, computer, and control engineers
- Electrical and power engineer
- Research chemist
- Process engineer
- Design and development engineer
- Construction engineer
- Shift engineers.

Although an individual for each of these professional skills need not be present for the startup, each area of expertise should be addressed. The presence of vendor representatives can be very beneficial to the startup team during an initial startup of combustion systems and gas turbines.

6.7. SUMMARY

Although plant startup occupies only a small fraction of the total life cycle of a plant, this is the period when most process safety accidents occur. The operations department has the lead role in startups. The maintenance department's role of providing temporary piping, emergency repairs, monitoring, and general assistance is important to the safety of the startup.

Planning and scheduling are the keys to safe and efficient startups. Each detail should be planned and scheduled, and open lines of communication should exist among the various disciplines and groups of the startup team. The procurement of quality raw materials, catalysts, and initiators is an important factor in initial startup. These materials are charged into the system for the first time, and it is during this time that any process safety problems will become apparent.

The surest way to guarantee a safe and efficient startup is to have operators and maintenance personnel who are experienced, well trained, and familiar with the process. Changes made during startup that do not invoke management of change procedures should be documented by operations and maintenance personnel, and appropriate changes should be made to the drawings; the operating manual; the startup, normal operating, and shutdown procedures; and to maintenance schedules and procedures. During startup, operations and maintenance personnel assume the lead role for maintaining process safety.

6.8. REFERENCES

1. *Large Property Damage Losses in the Hydrocarbon-Chemical Industries.* Marsh & McLennan, 14th edition, New York, NY, 1992.
2. Matley, Jay. "Keys to Successful Plant Startup." *Chemical Engineering*, September 8, 1969.
3. Holroyd, R. "Ultra Large Single Stream Chemical Plants: Their Advantage and Disadvantage." *Chemical Engineering Progress*, August 5, 1967.
4. Finneran, J. A., Sweeney, N. J., and Hutchinson, T. G. "Startup Performance of Large Ammonia Plants." *Chemical Engineering Progress*, August 1968.
5. Center for Chemical Process Safety (CCPS). *Guidelines for Technical Management of Chemical Process Safety*, American Institute of Chemical Engineers, New York, 1989.
6. Center for Chemical Process Safety (CCPS). *Plant Guidelines for Technical Management of Chemical Process Safety*, American Institute of Chemical Engineers, New York, 1992.
7. Lees, Frank P. *Loss Prevention in the Process Industries* (Volumes 1 and 2). Butterworths, London, 1983.
8. Horsley, D. M. C. and Parkinson, J. S. (Editors). *Process Plant Commissioning - A User's Guide.* Institution of Chemical Engineers, London, 1990.
9. Perry, Robert H., Green, Don W. and Maloney, James O. (Editors). *Perry's Chemical Engineer's Handbook*, Sixth Edition. McGraw-Hill Book Company, New York, 1984.
10. Sax, N. Irving. *Dangerous Properties of Industrial Materials*, Sixth Edition. Van Nostrand Reinhold Company, New York, 1984.
11. Baumeister, Theodore, Avallone, Eugene A., and Baumeister III, Theodore (Editors). *Mark's Standard Handbook for Mechanical Engineers*, Eighth Edition. McGraw-Hill Book Company, New York, 1978.
12. Natural Gasoline Association of America *Engineering Data Book*, Seventh Edition, 1957.

7

OPERATION

The long-term phases of a plant are operation and maintenance. The durations of the preceding phases of a plant's life cycle are finite, whereas the duration of the operation and maintenance phases is determined almost solely by economics.

The purpose of this chapter is to increase the supervisor's understanding of the elements of process safety that need to be applied during the operations phase. The roles and responsibilities of the first- and second-line supervisors when applying these process safety elements are defined, and methods of applying process safety elements are offered. This chapter covers the following topics: routine, nonroutine, and emergency operations; management of change; protective systems; training; incident investigation; human factors; and audits, inspections, and compliance reviews.

7.1. ROLES AND RESPONSIBILITIES

Although both the operations and maintenance departments are responsible for keeping the plant operating safely, the operations department has the primary responsibility for monitoring and controlling the chemical process and operating conditions that influence the safe production of specification product. Operations personnel will routinely start the process up and shut the process down. Operations staff must constantly monitor all systems to make sure that the process equipment is working as designed and to determine when adjustments and repairs are needed. Operations personnel will initiate the repair process and coordinate with the maintenance department to make the unit available for repairs. Because operations staff must know the process and equipment, safe operation of the plant requires operations staff training and refresher training.

Site and department management usually set policies and standards that govern how the plant is operated. Procedures are developed for the purposes of carrying out the policies and applying the standards. Depending on the local management culture, procedures can be developed by site management, department management, first- and second-line supervisors, or all of these. The first- and second-line supervisors who are directly responsible for carrying out the procedures, will usually be the first to recognize the need for additional procedures or changes to existing procedures.

The development of programs for implementing process safety management regulations is an example of how these roles work. The operations manager takes regulatory requirements and company and industry standards into account when developing a site-specific policy for specific elements of managing process safety. The operations superintendent then takes these site-specific policies and, with the help of first- and second-line supervisors, develops specific procedures for complying with the policies. Finally, the first- and second-line supervisors develop training requirements, arrange for the training, evaluate the effectiveness of the training, execute the procedures, and provide feedback to the superintendent about the effectiveness of the procedures and implementation of the program.

Operations personnel have a key role in process safety management. The first- and second-line supervisors are in the best position to enforce the complete and proper execution of the operating procedures. The need to monitor the execution of operating procedures is minimal when well-trained, knowledgeable, motivated, and alert operations staff are working. But the need for oversight is always there. Without vigilance, personnel can drift from the complete and proper execution of procedures, opening the door to a process safety incident.

7.2. ROUTINE OPERATIONS

Routine operations are the economic foundation of the plant. The objective is to routinely and safely produce specification product. A key objective during routine operations is to ensure the continuity of the process by maintaining specified operating conditions, constantly following operating procedures, and keeping an alert awareness of safety issues.

It is human nature that people become comfortable with the hazards to which they are routinely and regularly exposed. This can be illustrated by people's response to rain while driving. They slow down and proceed with more caution. But as the situation continues, they become more comfortable and increase their speed. The hazard, however, has not decreased. Supervisors must recognize that this behavior will occur in routine plant operations. The challenge is how to make personnel recognize and respect the hazard instead of becoming comfortable with it.

7.2.1. Operating within Process and Equipment Limits

Ensuring that a chemical process stays within safe operating limits requires the constant monitoring of operating variables and equipment performance by collecting process and equipment data, evaluating the data, and acting appropriately. The operating personnel analyzing the data should spot any potential safety implications. They should be able to clearly identify a situation where the process or equipment has strayed away from the safe operating limit and toward potentially unsafe operation.

Process operators need to know key historical information about the process and where to find it. Historical information about the design basis and the operating procedures is important to operational integrity. The reason behind a certain design feature or an operating procedure may be forgotten in time, leading people to change the design or disregard an operating procedure or limit. The routine operation of field storage tanks associated with a chemical manufacturing process illustrates this. All atmospheric pressure tanks have low design pressure limits. These limits are low enough that seemingly minor actions can overpressure the tank and cause its failure. An action as simple as nitrogen blowing a line into the tank can cause an internal overpressure and rupture a seam. Another example is steaming a tank during heavy rain, condensing the steam, leading to the collapse of the tank because of the internal vacuum.

Among the different types of routine operations, a steady-state, continuous process probably has the highest potential for the operators' loss of alertness and drift away from proper design procedures. Since concentrated operator activity is generally less than in a batch or cyclical process, the risk of operator mental drift is increased. First- and second-line operations supervisors have an important role in preventing this. For example, plant procedures sometimes allow for the disabling of certain alarms, but only under specific, approved conditions that include taking additional precautions. However, actual operating practices can sometimes, because of a lack of training and regular auditing by supervisors, deteriorate to disregarding alarms, or occasionally to disabling alarms critical to safety.

Managing routine operations that involve obvious hazards such as high temperatures and pressures is easier because the obviousness of the hazards makes people involved in these operations more aware. The supervisors' challenge becomes greater for more mundane operations—operations where awareness of the hazard can fade. An example incident is the unfortunate storage tank explosion that happened in an ARCO facility, July 1991.

Figure 7-1 offers a simplified diagram of the ARCO storage system. The incident involved an explosion of a large tank of liquid wastes, composed largely of water. Some flammable organic compounds, immiscible in water were also part of the contents. Some wastes could decompose to release oxygen. This characteristic was known, and two key design features were

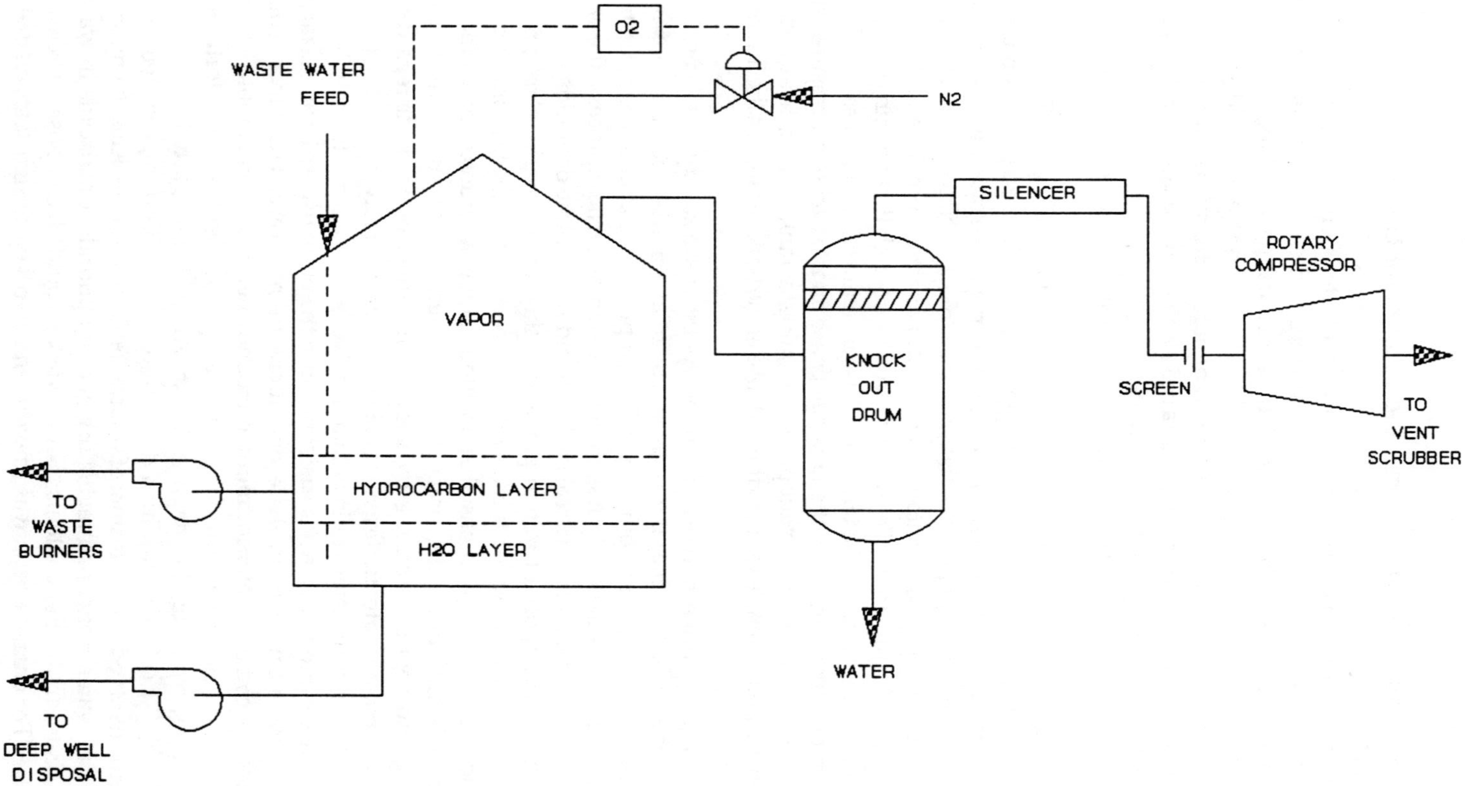

FIGURE 7-1. Storage system diagram.

included in the system: an on-stream oxygen analyzer for the tank vapor space and a nitrogen makeup to sweep the tank vapor space of oxygen. The investigators found that the oxygen analyzer had failed and was showing a constant low and safe oxygen level. The nitrogen sweep flow, controlled by the oxygen analysis, was essentially zero. In reality, the tank oxygen level had increased to result in a flammable vapor space, which ignited.

Among several lessons that can be learned from this incident, some clearly apply to the operations department. These are:

- The need to know and understand certain key design features and the reasons for them.
- The need to appreciate the limits of the design features. For example, know how the system will fail.
- The need to be operationally alert to a routine that is "too routine."
- The need to build reasonable backup checks into the operation.

These four lessons offer areas of focus for the supervisor because they are basic and can apply to virtually any process operation. How can the supervisor focus on these areas? The simple system of a conventional, fixed-roof, atmospheric pressure tank storing a flammable liquid can be used to illustrate these focus points. Preventing ignition by a nitrogen purge is designed into the system so that the vapor space will be nonflammable. The supervisor can develop and regularly teach a set of basic system principles. Tables 7-1 and 7-2 show such a written set of principles for this system. The principles for a specific, installed system would probably be shorter and simpler. The few key principles, however, should be known and regularly retaught.

A checklist may be helpful or even essential for monitoring the key features of the operation of a system. The supervisor should understand that sets of principles cannot be directly applied to an actual system or operation. Procedures and checklists are the bridge from principles to actual operation. Figure 7-2 illustrates a checklist that can be derived from the principles shown in Tables 7-1 and 7-2.

Continued and regularly refreshed awareness is critical to avoiding catastrophic incidents. Increasing process automation makes such awareness even more important. Supervisors should take a leadership role in ensuring continued awareness about maintaining operating procedures, monitoring process and equipment data, maintaining safe practices, and ensuring that all plant procedures and practices are followed diligently.

7.2.2. Written Procedures

An important role of supervisors is to ensure that the actual operating practices match the written operating procedures. Incidents are often caused by a deterioration of operating practices, such as the deactivation of alarm switches. For example, in one incident, a level alarm switch had slipped down

TABLE 7-1
Vapor Space Ignition Prevention Principles

FIXED ROOF TANK, INERT GAS BLANKETED

Type Storage: *Conventional Fixed-Roof Atmospheric Tank*
Vapor Space: *Nonflammable*
Method: *Inert Gas Blanketed*
Principles: *Operation*

Number	Principle
1	Application of the general operation principles.
2	Regular visual verification of the functioning of the inerting system. (1) Gas makeup system. (2) Overpressure relief devices open?
3	General oversight to confirm that regular maintenance inspections and preventive maintenance are being done as intended.
4	Regular operational visual checks of hatches, openings, air ingress opportunities.
5	Standard procedures existing and periodic training for: (1) Response to alarms/deviations. (2) Hatch gauging and sampling. (3) Pigging/purging/blowing operations. (4) Avoiding contamination, paying particular attention to interconnected vent systems.
6	Accuracy checks on analyzers.
7	Periodic secondary checks on actual vapor space oxygen content.
8	Regular operational visual checks on flame arresters.

NOTE: The **primary** barrier to ignition in this system is the effectiveness of the vapor space inerting system. A key operations consideration is the functioning of the inerting system.

its support into a sump and was alarming every few minutes as the sump pump cycled on and off. The operator taped down the "acknowledge" button on the console to stop the incessant alarm, thereby deactivating all audible alarms. The operator later failed to notice a high-temperature alarm light, an oversight that eventually led to a large, toxic vapor cloud release. Such incidents can be avoided by the continued vigilance of supervisors, who should ensure that operating procedures are not compromised. This involves monitoring personnel performance to ensure their adherence to the procedures. More important, such vigilance involves teaching and regular refresher training that emphasizes the importance of the procedures.

Supervisors often inherit existing written procedures for a process. That is, procedures written for an existing process or created by the project team in the design and installation of a new process. The supervisor may have been involved in this latter case.

TABLE 7-2
General Ignition Prevention Principles

ATMOSPHERIC PRESSURE FLAMMABLE LIQUIDS STORAGE SYSTEMS

Number	Principle
1	Regular operational visual check of:
	(1) Ignition prevention devices and equipment: appearance; functioning; obvious damage.
	(2) Electrical systems: appearance; obvious damage.
	(3) Electrical grounding systems: obvious damage.
	(4) Piping systems: appearance; obvious damage; correct lineup.
	(5) Openings and connections into tank vapor space: appearance; proper position and lineup; obvious damage.
2	Storage area controls on fixed ignition sources.
3	Storage area hot work controls and permit system.
4	Storage area administrative controls on other mobile ignition sources.

Tank No.: _____ Contents: _____

****

Note: **This tank is intended to be purged with nitrogen gas to keep the tank vapor space in a nonflammable condition. The focus of daily checks is on the purge system and oxygen content of the vapor space.**

****

Date: _____ Time: _____

Person making this check: _____

1. *Nitrogen system.*
____ a. Nitrogen flowing? **Yes - No**
____ b. Indicated flow (must be over 5 scfm) _____ scfm

2. *Tank hatches and openings.*
____ a. Pressure-vacuum manhole cover closed? **Yes - No**
____ b. Sample hatch closed and secured? **Yes - No**

3. *Check lineup to area scrubber. All three tank vent line valves open and locked open?* **Yes - No**
____

4. *Vapor space oxygen analyzer.*
____ a. Analyzer functioning? **Yes - No**
____ b. Analyzer field oxygen indication same as indicated on control room indicator? **Yes - No**
____ c. High oxygen alarm light and buzzer: Do they work? **Yes - No**
____ d. Analyzer automatic zero gas and spanning functioning? **Yes - No**
____ e. Date of last calibration and alarm set-point functional check (should not be more than one month ago). _____
____ f. Date of last field check of vapor space oxygen content (should not be more than three months ago). _____
____ g. Grounding connections: Check five grounding straps at tank base circumference. In good condition? **Yes - No**

****

Note here the immediate corrective actions for any deviations discovered during this checkup.

FIGURE 7.2. Checklist for ingnition prevention.

The supervisor has a continuing role updating, correcting, and enhancing the written procedures. This role is illustrated by a series of questions that the supervisor regularly addresses.

- Do written operating procedures exist?
- Do the members of the operating staff know that the procedures exist and use them?
- Are actual operating practices the same as those specified in the procedures?
- Are the procedures adequate? Correct?
- Do any of the procedures need to be enhanced (e.g., to be supplemented with checklists)?

When the answer to any of these questions is not satisfactory, the supervisor can start corrective action. In fact, the supervisor may be the only person in a position to recognize the need for corrective action. Not responding to a recognized need for correction creates a trap. A continued practice of *almost* following the procedures will eventually result in operating practices that can so deviate from the correct practice that an incident results. Not keeping written procedures current with actual operating procedures will result in written procedures that cannot be safely followed for the process.

A supervisor's actions to correct procedural problems usually fall into the following areas.

- Initial training and refresher training in the procedures
- Managing change of procedures
- Making key procedures more clear by enhancing them

Training and managing change are discussed in other sections of this *Guidelines.*

The supervisor's role in enhancing key procedures is important. Complete operating procedures need to be just that—complete, but completeness can make key procedures hard to find. Checklists are a way to enhance key procedures by making them brief and easy to extract from lengthy material. Some guidelines for enhancing and creating checklists follow.

1. Use checklists for key operational steps where:
   a. Improper execution of the operating step or procedure can directly lead to a release of a toxic or flammable material.
   b. The operating step or procedure is very complex.
2. Keep the checklist as simple and short as possible and exact in its meaning.

In summary, the supervisor's role regarding written operating procedures is to see that they represent reality and that actual operating practices are as specified in the procedures—always.

7.2.3. Communication

The likelihood of preventing an accident increases with the effective transfer of information. Communication among all plant personnel is critical, and supervisors are essential to these communications. Supervisors should provide information, see that the operations staff gets needed information and that they are communicating. The operational integrity of the plant depends on the frequency and quality of these communications.

The special position of the operations supervisors makes them privy to an enormous amount of information about the process safety status of the plant. To ensure safe operations, supervisors simply must communicate effectively to widely varied plant groups.

Supervisors need to recognize an important and powerful characteristic of communication and human nature. Things perceived to be important in an organization are not necessarily what is written in statements of principles, policies, and procedures—it is what the boss says and says regularly. The axiom is: The things that are important are the things that the next tier of supervision wants to talk about. Process safety-related issues (such as key operating procedures) decrease in perceived importance unless the supervisor directly, regularly, and plainly communicates their importance.

The opportunities for formally communicating process safety issues depend on the individual organization. Example opportunities are:

- *Safety meetings.* Persons involved in these are usually expecting to communicate about safety issues—usually personnel safety issues. These offer an excellent opportunity to discuss process safety issues.

- *Production scheduling meetings.* The safety content of these meetings will depend on the culture of the organization. Whatever the local culture, these meetings offer a powerful opportunity for the supervisor to discuss process safety, particularly if process safety is openly equated with operating or manufacturing safety.

- *Operations–maintenance planning meetings.* These meetings probably have the greatest potential for preventing process safety incidents. The very nature of planning the operations–maintenance interface and coordination involves process safety. Supervisors, both operations and maintenance, can use these opportunities to be sure that the meetings go beyond the mechanics of operations–maintenance planning. Supervisors are in a position to specifically communicate process safety issues once the mechanics are planned.

Communication implies a two-way flow of information, up and down the management hierarchy. Whatever the organizational structure, each level of management and supervision should convey sufficient information, which should at least include:

- A specific description of the work expected
- How the work should be done
- How the results of the work should be conveyed, and to whom

Finally, the communicators, whoever they are, should spend the time necessary to ensure that the communication is mutually understood. This last step is very important to operations supervisors. Proper communication is necessary for safely managing minor process changes. Minor changes that do not trigger the established management of change procedures require especially clear, specific communication because they are not subject to the checks that are part of established procedures. The supervisor has a key role in this communication.

Exchanging information with the technical services department also gives important support to the plant by developing improved production schemes and by improving the instrumentation, control, and safety of the process and equipment. To conduct their activities, the technical services staff relies on the operations and maintenance departments to provide them with critical production information. In many plants, this is accomplished through regularly scheduled exchange meetings among the technical services, operations, and maintenance departments.

Operational irregularities that may lead to process safety problems can be resolved in these interdepartmental communication exchanges. For example, in an ethylene purification plant, cold dimethyl formamide was used to absorb contaminant acetylene using cold service pumps to circulate the absorbent. A particular cold service circulating pump had a history of cavitating and was difficult to restart. One operator observed that this happened when a small decline in pressure occurred (but still within the published acceptable tolerance of the control system) in the absorber column. During a meeting with technical services, the first-line supervisor reported this observation. The technical services department found that when the absorber pressure dropped, it resulted in a lower pressure at the pump suction than the NPSH (net positive suction head) required by the pump, causing the pump to lose suction. Because it was in cold service, once it lost flow, the pump heated up and cavitated. The control tolerance of the pump instrumentation was tightened and the problem was solved. Without the type of communication illustrated here, solution of the problem would probably have been delayed, possibly to the point where a hazardous incident occurred.

Some companies operate the same or similar processes at different locations. An example is ethylene manufacture and purification. Companies with such multiple processes usually have established a method for regular, formal communication among the separate sites, such as technical exchange meetings. A powerful tool the supervisor has is to establish and maintain regular informal communication with counterparts at other manufacturing sites.

These interplant communications can be used to share information such as the following.

- Has the site experienced problems with a distinct portion of the process? How did personnel resolve it?
- Has the site experienced any process safety incidents in the process? What were the root causes? How were they corrected?
- How does the site conduct operator refresher training? What is emphasized in the refresher training?

This type of informal communication can be a powerful tool for preventing incidents and for sharing good operational "how-to."

Informal communication is difficult to measure or evaluate. Experienced operators generally agree that safe operations depend more on informal than on formal communications. Good leadership by supervisors is often characterized by their ability to promote an atmosphere conducive to informal communication and to show their respect for the knowledge and skills of those they supervise.

7.2.4. Communication During Shift Changes

The continuous operation of manufacturing processes requires multiple shifts of operating personnel. When operating personnel change shifts, the process is literally "handed over" to the next operations work team. Ongoing maintenance activities may require that shift change communications also occur between the maintenance work teams. The outgoing shift needs to communicate necessary information so that accountability can be properly transferred. The incoming shift should be sure that the necessary information has been communicated and understood so that they can properly assume accountability. This needed communication is not likely to be successful if left to chance. First- and second-line supervisors have the key leadership roles during this exchange of information. Also, supervisors are the persons who should ensure that other key operations and maintenance personnel exchange information as needed.

Actual methods and procedures for changing shifts are generally established by the local plant. Because of the potential consequences of "missed communication," certain minimum shift change practices are needed that will help ensure consistent process "hand over." Table 7-3 shows some critical issues that should be communicated during the shift change.

It is possible for vigilance to temporarily relax during the shift change because each work team may think the other team is responsible. Supervisors should guard against these less vigilant periods by making sure that both teams understand exactly when their responsibility ends or starts. Another problem that departing shift supervisors need to be aware of is that as it gets closer and closer to shift change time, operators may become reluctant to start

---

TABLE 7-3
Critical Issues That Should Be Communicated during Shift Change

1. The status of the plant, covering all aspects of operations and maintenance, should be communicated to the incoming shift. These include:
   —Current state of the unit
   —Temporary operations
   —Existing abnormal situations
   —Maintenance in progress

2. All process upsets or excursions that occurred during the departing shift should be discussed. Any corrective action taken should be described.

3. If a corrective action is in progress, the departing crew should explain the need for it and any emergency response actions activated.

4. Communication should occur about maintenance work permits in progress. The maintenance work team should transfer accountability about the work to the incoming maintenance work team.

5. Maintenance needs that should be addressed by the incoming shift should be communicated.

6. Any safety interlocks out of service, the reason for their being out of service, and the maintenance status of the interlocks should be addressed. Also, a description of any special measures needed because the safety interlocks are out of service should be given.

7. Any incidents or events that occurred during the shift should be communicated.

8. Any problems with instrumentation, controls, or utilities should be communicated.

---

an activity that they cannot complete before they are relieved. If the activity does not affect the process safety of the unit, it may be held over or delayed until the next shift. Critical activities that could cause a hazardous situation if not attended to immediately, should not be delayed, no matter how close it may be to shift change time.

Inadequate communication among operations and maintenance personnel and their supervisors during shift changes can lead to potentially hazardous situations. The following example illustrates the results of inadequate communication at such a time. In a multiproduct barge-loading facility, several products, including benzene and acrylonitrile, were loaded through dedicated lines. The benzene loading line had a block valve that was leaking and needed to be replaced. The line was emptied and decontaminated during the evening and graveyard shifts. During shift change, an operator improperly pointed out the acrylonitrile line to the supervisor as the decontaminated line. When the maintenance team reported to change out the valve, the day shift supervisor started them working on the acrylonitrile line rather than on the benzene line. When the flange on the valve was loosened, acrylonitrile sprayed out on the maintenance personnel who were not wearing protective clothing for working on acrylonitrile. Both maintenance technicians suffered serious chemical burns from the acrylonitrile. A key communication principle, ensur-

Factual description:

Two work permits were issued for maintenance work on a spare pump and a relief valve associated with the spare pump. Maintenance had electrically isolated the pump. Also, the pump relief valve was removed for overhaul and the open end was blanked. During the next shift, the main pump failed. Operations personnel may not have known that the relief valve was missing and may have recommissioned and started up the spare pump. The blind flange was probably not tight and the propane leaked past it, ignited, and exploded.

Contributing causes	Process safety concept
1. The permit-to-work system and shift handovers of work in progress were inadequate. Procedures should have required the departing shift to make sure that the incoming shift knew about the status of all work in progress. This would have eliminated the possibility of recommissioning the spare pump.	Accountability, Operating procedures. Process and equipment integrity
2. The diesel fire pumps were on a manual mode, which inhibited the operability of the fire water system.	Equipment integrity, Emergency planning
3. Drills and training for emergencies did not take place as frequently as specified in the procedures.	Adherence to procedures, Emergency planning
4. Lock, tag, and try procedures should have been adhered to.	Adherence to maintenance procedures

FIGURE 7-3. Piper Alpha incident

ing mutual understanding, was missed. This example of poor informal communication during a shift change also reveals other safety issues, such as the absence of a work permit and the failure of the maintenance team to follow safe work practices when opening the flange. Multiple safety protection systems can be rendered useless if personnel assume that one layer of protection is sufficient and bypass the remaining layers.

The Piper Alpha accident, [1] described in Figure 7-3, is an infamous incident in the petrochemical industry in which lack of communication between two shifts was a contributing cause. How can operations supervisors ensure safe and successful shift change communications? The supervisor at least needs to understand the local shift change communication procedure and ensure that the operations participants in the procedure understand it and follow it. This assumes, of course, that a local procedure exists. The rest of this section will cover some key characteristics of a shift change communication procedure.

The local unit or shift supervisor can either enhance existing procedures for specific unit application or develop a unit procedure when a local shift change procedure does not exist. Local operations unit supervision and maintenance work team supervision should review the specific needs and conditions before developing a unit or team shift change procedure. This step can be accomplished short-term by a small team of persons involved in the shift change process. The development team should at least address the following:

1. State the objective for the development team.
2. State the objective for the shift change and for the individual participants involved in shift changes.
3. Identify the key unit specific details that must be addressed in each shift change communication.
4. If appropriate for the local situation, define a "pattern" or "format" for shift change.
5. Consider the creation and use of shift change checklists as a means to help assure thoroughness and consistency.
6. Decide how frequently the unit shift change procedure should be reviewed:

This overall development process is illustrated in Figure 7-4.

The shift change procedure may include some or all of the elements of the example procedural flow chart shown in Figure 7-5. For local use, the procedure can be carried out with a checklist. Figure 7-6 shows an example checklist.

The number one priority during shift change is for the departing team to communicate the status of the process and equipment to the incoming team. The shift supervisors should make sure that the exchange of necessary information is taking place as required by procedures and common sense. Depending on the organizational culture, the shift change communication can be accomplished solely by the individual participants, by means of a group exchange, or a combination of both. Whatever the method of communication, the incoming shift should clearly understand the status of the process and equipment. There should be sufficient overlap of shifts to ensure proper monitoring of the process during the change.

7.2.5. Special Safety Considerations of Batch Processes

Specialty chemicals and pharmaceutical manufacturing processes usually require batch processing. In other plants, batch processing usually exists within or on the fringes of continuous processes. Batch processes consist of a combination of several discrete tasks or operations that must be performed sequentially to produce products up to specification and to maintain process and equipment integrity. Thus, the effectiveness of a batch operation will depend on four factors [2]:

- Analysis of individual operations.
- Coordination of operations.
- Overall production sequence for one product.
- Overall production strategy for multiple products.

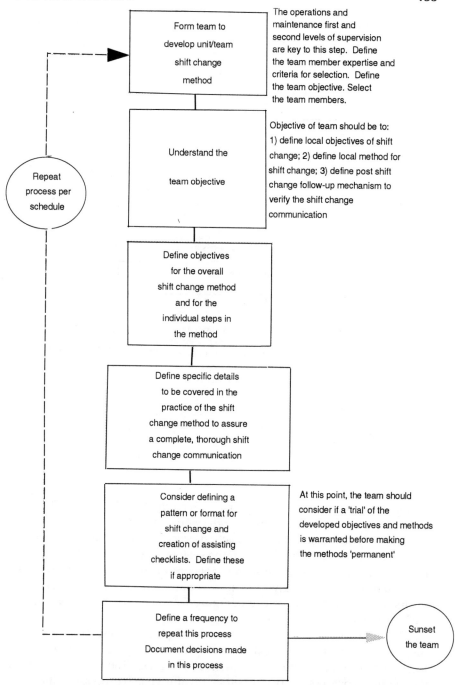

FIGURE 7-4. Basic process: Shift change. (A generic process to develop shift change method.)

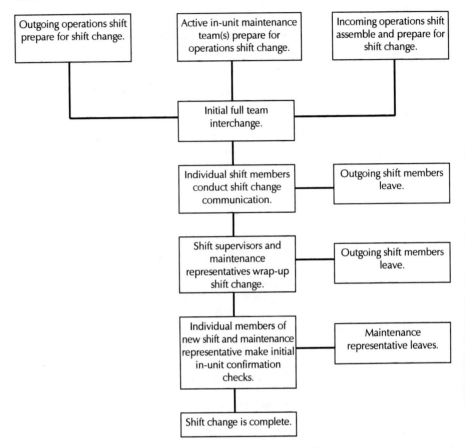

FIGURE 7-5. Example shift change procedure.

Operations personnel are highly involved in all of the above. This active operations participation requires a high degree of awareness.

A key characteristic of batch reaction processes is that all of one or more of the reactants is usually present in the system at the start of the processing. If the reaction rate of such a system is not controlled, the reactant available in the system can become involved in a runaway reaction with catastrophic results. The incident[3] described in Figure 7-7 illustrates this characteristic of batch processes.

The safety of a typical batch process depends on consistent sequencing of the process steps. Modern-design batch processes typically will be on computer control and have built-in sequential interlocks. Table 7-4 shows examples of critical process safety issues that can affect batch processes. Since batch processes consist of sequential tasks that must be completed one after the other, it is quite common to use checklists. These checklists may contain the

Date: _____

	Outgoing	Incoming
Supervisor	_____	_____
Reaction Operator	_____	_____
Reaction Outside Operator	_____	_____
Purification Operator	_____	_____
Purification Outside Operator	_____	_____

****

Preparation

1. Any maintenance work in progress?   YES   NO

    If yes, list on other side and be sure that it is fully described in shift log book.

    If yes, enter name(s) of maintenance contact(s) on other side.

2. Any alarms, interlocks, automatic shutdown systems out of service?   YES   NO

    If yes, list on other side and be sure that these are fully described in shift log book.

3. Any temporary operations in progress?   YES   NO

    If yes, list on other side and be sure that these are fulling described in shift log book or in the
Temporary Operating Procedures.

****

Shift Change

    1. Each outgoing operator exchanged with incoming.

_____   Reaction Operator

_____   Reaction Outside Operator

_____   Purification Operator

_____   Purification Outside Operator

_____   2. Covered status of unit.

_____   3. Covered status of maintenance activities.

_____   4. Covered temporary operations.

_____   5. Covered status of alarms, interlocks, shutdown systems.

Outgoing Supervisor:_____   Time:_____

Incoming Supervisor:_____   Time:_____

****

Shift Change Follow-up

    1. Read the shift log book back to last shift worked.

_____   Supervisor.

_____   Reaction Operator.

_____   Reaction Outside Operator.

_____   Purification Operator.

_____   Purification Outside Operator.

    2. Checked on maintenance activities.

_____   Supervisor.

_____   Reaction Outside Operator.

_____   Purification Outside Operator.

_____   3. Checked with the maintenance contact(s).

Supervisor: _____   Time: _____

****

Note: File this completed checklist in Shift Log Notebook.

FIGURE 7-6. Shift change checklist—ethylene oxide unit.

Factual description:

Figure 7-8 shows a batch reaction system. A batch of glycerol was placed in the reactor and circulated through a heat exchanger that could act as both a heater and a cooler. Initially it was used as a heater, and when the temperature reached 115°C, the addition of ethylene oxide began. The reaction was exothermic and the exchanger was now used as a cooler.

The ethylene oxide pump could not be started unless:

1. The circulation pump was running.
2. The temperature was above 115°C, because otherwise the ethylene oxide would not react.
3. The temperature was below 125°C, because otherwise the reaction was too fast.

Despite these precautions, an explosion occurred. One day, when ethylene oxide addition began, the pressure in the reactor rose. This showed that the ethylene oxide was not reacting. The operator decided that perhaps the temperature point was reading low, or perhaps a bit more heat was required to start the reaction, so he adjusted the trip setting and allowed the indicated temperature to rise to 200°C. Still the pressure did not fall.

He then suspected that his theory might be wrong. Perhaps he had forgotten to open the valve at the base of the reactor? He found it shut and opened it. Three tons of unreacted ethylene oxide, with the glycerol, passed through the heater and catalyser and a violent uncontrolled reaction occurred. The reactor burst and the escaping gases exploded. Two men were injured. One, 160 meters away, was hit by flying debris, and the other was blown off the top of a tanker.

Although the indicated temperature had risen, the temperature of the reactor's contents had not. Pump J2, running with a closed suction valve, got hot and the heat affected the temperature point, which was close to the pump.

Contributing causes	Process safety concept
1. The explosion was due to an operator forgetting to open a valve. It was not due to lack of knowledge, training, or instructions, but was another of those mistakes that even well-trained, well-motivated, capable people make from time to time.	Refresher training and awareness.
2. If the operator had not opened the valve when he found it shut, the explosion could have been avoided. However, it is hard to blame him. His action was instinctive. What would you do if you found undone something you should have done some time ago?	Training (e.g., What-if simulation, Process simulator)
3. The explosion was due to failure to heed warning signs. The high pressure in the reactor was an early warning but the operator had another theory to explain it. He stuck to this theory until the evidence against it was overwhelming. This is known as a mind-set or tunnel vision. The other temperature points would have helped the operator to diagnose the trouble. But he did not look at them. He probably thought there was no point in doing so. All the temperature points were bound to read the same. The need for checking one reading by another should have been covered in the operator's training.	Training procedures to ensure training achieves goals training.
4. The explosion was due to a failure to directly measure the property we wish to know. The temperature point was not measuring the temperature in the reactor but the temperature near the pump. This got hot because the pump was running with a closed suction valve. Similarly, the trip initiator on J2 showed that its motor was energized. It did not prove that there was a flow.	Design flow, Process hazards analysis.
5. The explosion occurred because key instruments were not kept in working order. The flow indicator and low flow alarm (FIA) were out of order. They often were, and the operators had found that the plant could be operated without them. If there is no flow, they thought, J2 will have stopped and this will stop J1.	Mechanical integrity, training, process safety awareness.
6. The operator should not have raised the trip setting, al though it did not cause the explosion. (However, he did try to use his i ntelligence and think why the reaction was not occurring. Unfortunately, he was wrong.)	Procedures, management of change.

FIGURE 7-7. Batch process incident[3].

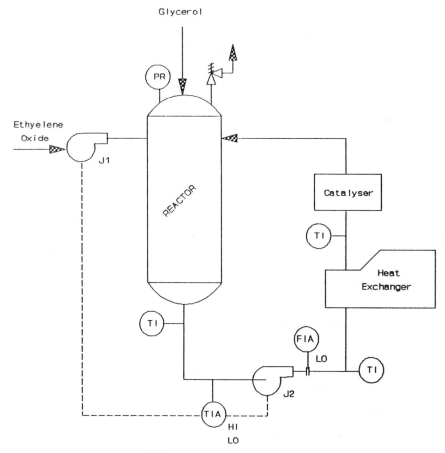

FIGURE 7-8. Reactor circulating system.

sequence of certain procedural steps and the consequences of not following those steps. In addition, the checklists may cover other safety issues. Table 7-5 shows some key elements of a typical batch checklist.

Batch control, no matter how complicated the process, has all of the characteristics of a recipe used to prepare a meal. It is a chronological sequence of events or actions. The chronology of events is recorded step by step and is called a batch sheet, which records the sequence of events, who is to carry them out, the level of supervision, and any deviations or comments. Figure 7-9 is an example of a batch sheet. The extent and importance of operator involvement in the process is quite clear from the activities shown in the table. It is the responsibility of the operations department to maintain records of the batch sheet. Also, these sheets should be reviewed periodically to find problems and correct discovered deficiencies.

TABLE 7-4
Examples of Critical Process Safety Issues That Can Affect Batch Processes

Batch process issues	Process safety concepts
1. All points in the process should be instantly accessible to any interlocks without interfaces for each device in the data highway.	Process and equipment integrity
2. Operators should be familiar with the control software and understand the implications of different statements in it.	Process knowledge and documentation, Procedures, Training
3. Indexed printouts of interlocks should be available.	Process knowledge, Accessibility of documentation
4. Strict rules should govern when and how safety interlocks can be changed or bypassed	Procedures, Process and equipment integrity
5. The control system should be validated before every startup.	Pre-startup safety review, Procedures
6. Maintain awareness of risks external to the process (e.g., feeding the wrong materials).	Process knowledge, Training
7. Avoid unwanted reactant accumulation.	Process knowledge, Procedures, Training
8. Reduce or eliminate the possibility of formation of dust clouds, especially for agitators and powder-feed systems.	Process hazards analysis, Procedures, Training
9. Reduce or eliminate entry of air when reactants are added.	Process hazards analysis, Procedures, Training
10. Management of change procedures should be applied to all modifications, major or minor.	Management of change

---

In many plants, batch processes exist on the fringes of continuous processes. Thus, the continuity, smooth operation, and process integrity of the batch process affects the operations and process integrity of the continuous process. Since batch processes by nature consist of a series of sequential operations, the execution of one operation may depend critically on the previous operation or affect the next one. Thus, the sequencing of batch operations is critical, whether it is a pure batch process or a batch process existing within a continuous process.

The example described in Figure 7-10 illustrates the critical nature of batch sequencing. As explained in the figure, a process hazards analysis should not only have addressed the need for installing interlocks, but should also have recommended changes in operating procedures and training to address scenarios where the critical nature of the batch sequencing affects the overall integrity of the plant.

TABLE 7-5
Key Elements of a Typical Batch Checklist

1. Description of the operational steps.

2. Information on the operating limits, warnings, or consequences of deviating from the operational steps or operating limits.

3. Procedures for emergency actions.

4. Instructions to check if the previous batch has been dumped and the vessel has been isolated.

5. If the procedures require decontamination, instructions to check if decontamination procedures have been completed.

6. Instructions to check the batch reactant weigh tanks for the appropriate amount of reactants.

7. Instructions to make sure that the reactor agitator is turned on.

8. Instructions to check and make sure that vent valves and pressure control are operating adequately.

9. Instructions to check the cooling system.

10. On addition of reactants, a reminder to check to see that the reactant rate is adequate and that the agitator is functioning properly.

11. Instructions to make sure the temperature control profile is functioning properly and that the temperature is being maintained accordingly.

12. When reactor feed tanks are empty, a reminder to isolate the feed tanks from the reactor and continue the reaction until endpoint.

13. Instructions for dumping reactor contents to the receiving tank and shutting off the agitator.

Depending on the complexity of the plant and the automation involved, operator participation in beginning different types of activities can vary. In most batch processes, the level of operator participation is high. Thus, the proper practice of procedures is key to maintaining process and equipment integrity. Table 7-6 shows examples of some batch process activities that could lead to unwanted incidents.

In most batch processes, reaction rates are defined by the rate of reactant addition or the temperature of the reaction batch. To prevent runaway reactions, it is critical that the specified reaction rates be maintained. Reaction rates that are too high can result in dangerous runaway reactions. Reaction rates that are too low can result in an accumulation of reactants that simply await the conditions for a runaway reaction. It is therefore critical to maintain the reaction rate in a desired range by controlling the addition of the reactants and the temperature of the reaction batch. As shown in the example in Figure 7-11, operator awareness of safety critical steps (process knowledge and documentation) is essential to avoiding incidents. In the example shown, a process hazards analysis should have investigated the consequences of a reduced flow

EXTERNAL INSPECTION AND PREPARATION FOR CHARGING FEED	
Introductory Information	
1. Safety equipment/requirements: Hard hat, safety glasses with side shields, ear protection, rubber gloves and safety shoes. 2. Kettles are glass-lined and can be internally pressurized, externally heated, and fitted with a scrubber line to vent fumes.	

Steps	Key Points
1. Check with supervisor on status of kettle.	Supervisor will confirm status of kettle for charging and scrubber for operation.
2. Check all numbered lines, condition of valves, and gauges for leaks, corrosion, damage, etc.	If you find anything unusual, notify the supervisor, if not proceed to step 3.
3. Open scrubber valve #2 and pressure gauge valve #6.	(a) *Safety:* Open this valve from the *rear* of the kettle *only*. This valve is opened in a counterclockwise direction to ensure proper venting of kettle. (b) There should be **no** pressure on kettle. The gauge should read 0.
4. Go to emergency shut-off panel and pull out emergency-close button to operate air-activated bottom valve #1.	Emergency shut-off.
5. Close air actuated bottom valve #1.	(a) Bottom valve switch is located at bottom of back room stairway, to the right. (b) Push one, *top* button and you should get a red light indicating valve is *closed*. (c) Push two, *bottom* switches simultaneously and you should get a green light indicating the valve is *open*. *Caution:* If any of the above conditions are not met, stop here and notify supervisor.
6. Close valve #7a on drowning leg.	This is in case of valve #1 failure.
7. Push in emergency shut-off on floor panel.	Bottom valve #1 cannot be accidentally activated in this position.
8. Install gasket, close manway, and secure with minimum of three J-bolts.	Make sure manway cover and gasket and kettle surfaces are clean and in good condition.

FIGURE 7-9. Example batch sheet.

Factual description:
An incident caused by holding product for too long occurred in a batch process where a hydrolyser and a crystallizer were being used in series. The crystallizer was shut down because the agitator failed. While repairs were being made, the batch in the hydrolyser was held. Because of a fault in the hydrolyser temperature indicator and controller, the hydrolyser batch continued to be heated until it decomposed and exploded. The incident caused the death of one person and extensive damage to the plant.

Contributing causes	Process safety concept
1. When holding batches, batch temperature should be verified and monitored continuously.	Procedures, Training
2. A process hazards analysis should have investigated the consequences of holding the batch in the hydrolyser. The results should have been incorporated in the procedures, which would have described the length of time the batch could be held in the hydrolyser, any additional safety precautions needed, and exactly when the shutdown should begin.	Process hazards analysis, Operating procedures, Training
3. Process hazards analysis should also have specified the necessity of installing interlocks.	Process hazards analysis, Equipment integrity
4. If the decision was made to hold the hydrolyser batch for a longer period, it should have been approved by management of change procedures for temporary changes.	Management of change, Training

FIGURE 7-10. Critical batch sequencing.

of cooling water. The results of the process hazards analysis should have been incorporated into the operating procedures and training program.

In summary, the supervisor should sharply focus on batch procedures. The supervisor should realize that changes in the proper execution of procedures—even seemingly slight drift in the procedures—are, in fact, process changes. A process change by way of procedure drift can occur more easily than process changes by way of process hardware changes. Process changes by way of procedure drift can be insidious and almost invisible, but they are process changes none the less and can lead to dangerous consequences if not managed. The first and most important step in the supervisor's management of the procedures is to ensure a clear and regularly refreshed understanding by the operations team of the procedures and the consequences of deviating from the procedures.

7.2.6. Process Control Software

The trend in the chemical process industry is to move to computers for operations control instrumentation. The operator's view of the operating process becomes the window offered by the computer and the software. That

TABLE 7-6
Examples of Batch Process Activities That Require
a High Degree of Operator Awareness

1. Using the same equipment for products that may pose the risk of cross contamination (e.g., highly reactive chemicals, highly active pigments, different grades of polymers). Even the most efficient decontamination and cleaning procedures may not succeed in eliminating the risk of hazardous cross-contamination. Decontamination procedures should include a final verification step.

2. Charging reactors or vessels with reactants. A number of hazardous incidents have been caused by improper charging procedures. For example, powder being fed through a chute into a reactor may ignite from static electricity generated on insulated metal components of the plastic chute by the passage of the powder.

3. Monitoring and control of the reaction rate. The batch reaction rate is a function of the rate of addition of reactant and the temperature of the reaction batch. To keep the reaction rate within safe limits, the reactant addition rate and reaction batch temperature should be monitored carefully.

4. Changes in process chemistry. Batch process chemistry is susceptible to dangerous changes because of a loss of agitation, loss of cooling or heating, or the loss or addition of a reactant.

5. Strict adherence to batch recipes. Too much catalyst or reaction initiator can cause a runaway reaction. If a deviation from the batch recipe is made, the change should go through a hazard review, and the emergency response plan should be sufficient to cope with any emergencies.

6. Strict adherence to batch operating sequence. An incorrect operating sequence can lead to catastrophic incidents. For example, the addition of reactants without turning on the agitator can initiate a runaway reaction.

7. Monitoring of batch size. Batch size should be increased only after the appropriate management of change considerations, which should include the effects on safety-protective equipment (e.g., PSVs and PSEs may not be adequately sized to safely relieve the larger mass) as well as on downstream equipment and units. Downstream equipment and units may not have the capacity to receive the increased capacity. Also, before upstream units are started, downstream equipment should be checked to make sure they are ready to receive the batch.

Factual description:
In a process to produce an acetic acid salt, hexamethyl tetramine and cyanide were reacted. The reaction was proceeding normally when a malfunction caused a reduced flow of cooling water. The reactant addition rate was not lowered accordingly to compensate for the reduced flow of cooling water. Consequently, the pressure in the reactor increased to a point where the rupture disc released large quantities of cyanide vapor, exposing several workers to the cyanide. Favorable wind conditions averted more serious consequences.

Contributing causes	Process safety concept
1. Failure to maintain proper reaction temperature by reducing the reactor feed rate while correcting the malfunction that reduced the flow of cooling water.	Process knowledge and documentation, Operating procedures, Training
2. A process hazards analysis should have investigated the consequences of reduced water flow. The results of the process hazards analysis should have been incorporated in the operating procedures and training program accordingly.	Process hazards analysis

FIGURE 7-11. Cyanide incident.

window can be as narrow or wide, simple or complex as the power of the computer hardware allows and as the software designers have created. The creation and design of the window probably involved operations expertise during the design phase of the plant.

The supervisor should focus on two key long-term issues about the process control software.

- Is the operator's window into the process effective?
- Is the control software secure?

The effectiveness of the window will probably not be questioned unless someone (the supervisor) regularly questions it or, much worse, an accident happens because the window was not effective. The forcing of the question is important simply because we humans "make do" with what we have, even when the current state of affairs may not be best.

The supervisor can encourage this questioning by periodically getting the operations work team to explore questions like the following.

- What parts of my window do I routinely use?
- What parts do I seldom or never use? Why? Do these ever obstruct the useful parts of my window? How?
- What parts of my window must I have immediately and without inhibition or dilution when I am in an abnormal operating situation? Why? Do I have that immediate, undiluted, uninhibited access? If not, why not?

The work team's periodic answers to these questions can be a powerful tool to starting needed management of change procedures for keeping the operating window current and genuinely useful.

The process safety of a plant also depends on the security of the control software. Control software must be changed from time to time, and these changes should be treated as process changes. Proper management of change procedures should be followed before the changes are made, and the supervisor should verify that any control software changes have the proper authorization. Under no circumstances should the supervisor allow unauthorized software changes.

A plant usually has procedures covering the security of process control software. Procedures for changing software should be the same as or similar to management of process change procedures. The use of passwords and other security measures is similar to lockout/tagout procedures. For changes in the software, multiple password security is necessary to avoid error and inadvertent software changes. Data input also requires security measures, but not at the same level. A password should be required for data entry and, if possible, some check on the validity of the data should be made. For example, data can be compared by the software with a range of reasonable or past data to eliminate gross and possibly dangerous entries. A simple "Are you sure?"

query after the data entry step is vulnerable to operator mental drift and is easily ignored.

The real security of the software depends on the discipline of the persons who have access to the software or who can gain access, whether the plant has security procedures or not. The supervisor has an important role in maintaining that discipline, largely by maintaining an awareness of the issue among the work team. An important awareness issue is illustrated by comparing a piping change to a software change.

- Once completed, a piping change is tangible and can be seen; it can be questioned again if need be.
- Once completed, a software change cannot be readily seen; only the results of the changed software can be readily experienced.

7.3. NONROUTINE OPERATIONS

Nonroutine operations are characterized by infrequent practice. Nonroutine operations can be either planned and scheduled or can occur without an opportunity to schedule them. Examples of planned nonroutine operations follow.

- Periodically recharging catalyst to an operating homogeneous catalyzed reaction system
- Mixing batches of inhibitor solution that will satisfy several days of operation
- Making a final live steam purge of a chemical sludge recovery still at the end of a multiday recovery cycle
- Unloading delivered hazardous feed chemical into process feed tanks that hold multiweeks supply

The characteristic infrequency of planned nonroutine operations poses both a threat and a benefit to process safety. The threat results from treating the nonroutine as routine without being sure that the immediate understanding of the operation by the operator is correct. The benefit can be enjoyed by simply being disciplined to briefly refresh an understanding of the operation just before conducting the operation. The supervisor can influence the discipline to negate the threat and gain the benefit.

Two other important categories of nonroutine operation, abnormal operations and stand-by operations, normally are not scheduled. The supervisor can, however, plan for them. These two categories are discussed in the following sections.

7.3.1. Abnormal Operations

Abnormal operation can be characterized as any operation outside the upper and lower operating limits, which should normally be defined in the operating procedures. These limits have their roots in the safe upper and lower limits normally defined in the process technology information. Normal operating procedures will include information about the consequences of deviating from operating limits and information about how to recover from such a deviation. This latter information is key to the supervisor's planning for the abnormal.

Abnormal operation does not necessarily imply an immediate threat. The abnormal can, however, escalate to a threat if a proper response is not made. One type of abnormal operation is a temporary or experimental operation, planned and scheduled to be carried out beyond normal operating limits.

Experimental operations, by definition, involve process changes; therefore, these types of operations should go through the plant's management of change process. Key considerations that the operating team should clearly understand before beginning experimental operations are:

- The objective of the experimental operation.
- Procedures for the experimental operation, including the experimental operating limits and responses to deviations from those experimental limits. The affected operating and maintenance work teams should be trained in these procedures and understand them.
- The period during which the experimental operation will be in effect.
- New emergency planning and response needs that might result from the experimental operations.

The unexpected abnormal operation obviously is not scheduled but the supervisor can plan for it. For planning and preparation, the supervisor will need to know and work with the following key principles:

- The operating work team members need to know the basic operating responses to deviations from limits and where to quickly find the details for those responses. Regular training should be given to refresh the team's understanding of this key part of the operating procedures.
- The operating team members should be able to recognize an abnormal situation or condition when it occurs. Usually, well-designed process instrumentation and alarms serve this need. But also, unusual noises, vibrations, or smells should not be ignored by the team, and they should understand how to respond to these potential alerts.
- The supervisor and the operating work team should recognize the inherent tendency of people to go beyond acceptable operating limits to try to save a process, or in other words, the inertia against initiating a process shutdown.

Operating limits are usually thought of in terms of temperature, pressure, and flow (or quantity of material). Another important measure is time, particularly the time allowed for certain steps of a batch reaction. The Seveso incident, in Italy in 1976, is a classic example of a process time abnormality that resulted in a serious accident.

The process was a batch reaction to produce an herbicide and it included multiple process steps involving different chemicals in a batch reactor. Certain time limits were associated with each step. For certain reasons, the plant was shut down for the weekend in mid-batch and left at that stage—something never before done. Partly because of a series of other deviations from normal operating limits that occurred within the now idled total plant system, an exothermic reaction began in the batch reactor and eventually escalated to a runaway reaction and discharge of the reactor contents to the environment. To compound the situation, the runaway reaction under the abnormal conditions caused the formation of highly toxic dioxin.

Table 7-7 gives examples of abnormal conditions that can potentially escalate to catastrophic consequences. When unexpected abnormal operating conditions are detected, the first objective of the operators is to bring the plant back to a normal operational state. However, if the conditions deteriorate, a shutdown should be initiated. Decisions must be made quickly and reliably. Shutdown procedures should clearly state where efforts to recover from abnormal conditions should be abandoned and shutdown initiated.

In cases of extensive computer-controlled, automated processes, a key concern is losing automation. Three problems that can be encountered are:

- Are there enough manual controls to manually operate the process effectively?
- Does the operating work team practice enough to operate manually?
- Will the "gap" between normal operations and safe limits need to be widened for manual operation?

The supervisor should regularly refresh the work team about procedures. For example, some supervisors may schedule training sessions on how to operate units manually. Some parts of the process may be available for the team to practice manual operation. Process simulators are handy for refresher training in manual operation. In addition, the what-if games discussed in Section 7.7.2 can also be used.

Training for abnormal operations is more difficult. First, choices should be made about which of the many possible abnormal conditions need special focus. The increase of sophisticated process controls can present sudden and complex problems to the operator when abnormal conditions result from malfunctions in those controls. The operator should be able to return the process to normal conditions. Because such malfunctions are rare, the skills acquired through normal training may be forgotten. Thus, the operating work team should be regularly refreshed in key skills.

TABLE 7-7
Examples of Abnormal Operating Conditions That May Lead
to Potentially Catastrophic Incidents

Abnormal condition	Potential hazards
Elevated reaction temperature	Side reactions, runaway reactions
Low reaction temperature	Could cause a dangerous accumulation of reactants that can result in a runaway reaction when temperature is increased
Agitator failure	Runaway reactions, explosions
Loss of coolant	Runaway reactions, explosions
Fracture of an internal coil	Safety problems may occur if the fluid is incompatible with the contents or at a significantly different pressure than the contents
Incorrect reactants, incorrect reactant ratios, incorrect order of reactant addition	The potential hazard may vary from nothing to off-specification product or violent uncontrollable reaction or explosion
Charging too fast or too slowly	Charging too fast may cause high reaction rate, leading to overpressure and explosion
Inadequate dispersion of immiscible reactants	Uncontrollable reaction when the charge is dumped to the dump tank
Power failure	Could be disastrous for reactions that require significant cooling to maintain control
Sensor failure	Failure to track control parameters can lead to potential hazards
Incorrect instrumentation setting	Failure to track control parameters can lead to potential hazards
Blockage of safety relief valve	Could result in vessel overpressure leading to failures and releases of hazardous chemicals

7.3.2. Standby Operations

Standby operations are a special type of operation in which the process is kept operational without any forward feed (e.g., total reflux for a distillation process, reactor with hot gas circulating, etc.). To draw a parallel, standby operations are similar to idling an automobile engine. However, while an automobile can theoretically stay on idle indefinitely without causing hazards, chemical processes can stay on standby operations only for limited time.

Although no feed is added and no product is withdrawn, standby operations should conform to all of the process safety procedures required for normal operation. It is important to have good preplanning and procedures and a clear understanding of the limits, including time limits, of standby operation. Some processes (e.g., distillation, purifications, etc.), may be kept on standby for a long time without creating any hazards. A standby operation for a distillation column would require the column to be inventoried with materials. The distillation process would then be started up and the column temperature would be maintained at a constant level. The reboiler also would stay operational and the column would operate on total reflux. No feed is added nor are bottoms or overhead withdrawn. When it is established that the column will begin to produce forward product, the addition of feed begins and the withdrawal of bottoms and overhead starts. Standby operation and transition to normal operation require careful monitoring and correct responses at the appropriate time. Losing control of the process may cause problems ranging from the production of off-specification product to the release of hazardous materials.

In contrast to distillation and purification processes, reaction processes can be maintained on standby operation for only limited periods. For example, long periods of standby operation involving a catalytic reactor can damage the catalyst bed or cause hazards related to the reactive materials in the system. The operating procedures for the unit should describe the safe limits for holding the process, and it is important to follow these procedures strictly. Figure 7-12 describes an incident that illustrates some critical safety aspects of standby operations. The incident described could have been avoided if the limits of the standby operation had been clearly understood and followed by the work team.

Factual description:	
Propylene and ammonia were used to produce acrylonitrile in a batch reactor. During a standby operation, inert gas was circulated over the reactor bed while the instrumentation was being checked and downstream units were being readied. The standby operation continued for longer than usual, causing the reactor bed to overheat. Consequently, the introduction of feed to the reactor led to a faster rate of reaction than normal, and ultimately to a runaway reaction. The reactor overpressured and released a large quantity of reaction products, resulting in a serious fire.	
Contributing causes	**Process safety concept**
1. The limits of the standby operation were not clearly understood and followed by the work crew.	Process knowledge, Procedures, Training
2. A process hazards analysis should have identified the possibility and consequence of longer standby operations and recommended the installation of temperature alarm and shutdown devices.	Process hazards analysis, Process and equipment integrity
3. The decision to continue standby operation for a longer time should have been made only after appropriate management of change considerations for temporary change.	Management of change

FIGURE 7-12. Runaway reaction during standby operation.

The seemingly innocuous distillation standby operation can be entirely different if the material is reactive (e.g., a reactive monomer that can exothermically polymerize if incubated long enough at a high enough temperature). The key to safe standby operations is establishing limits for the standby, understanding those limits, and practicing them.

7.4. EMERGENCY OPERATIONS

The operations team will usually be the first to detect an emergency condition and therefore the first to begin to execute the initial activities of an emergency response plan. It is essential to maintain operator awareness about how to recognize emergency conditions and respond appropriately. The operators can probably visualize different emergency conditions that could occur under different operating conditions. An important tool for increasing team awareness is use of the so-called "war games" during slow times on shift. Many organizations go through these what-if types of scenarios of simulated emergency conditions. For example, the supervisor may have the operations work team go through a drill of an emergency shutdown of a furnace. The furnace would not actually be shut down; however, operators would talk through their respective activities and possibly go to the locations for those activities, as described in the emergency response plan. They would go through the motions of executing the steps described in the plan and communicate their actions to the team. Finally, the team would discuss the intended "secured" status of the unit or the plant after executing the plan. The supervisor can then lead the team in a critique of the simulation. Key points for the team to critique and understand are:

- How they performed in the simulation. Did the simulation follow the procedures and plan?
- Would following the procedures have resulted in successfully securing the unit?

Emergency conditions create a lot of stress. Also, an added danger is the instinct to try to save the process (i.e., continue to operate) and avoid a shutdown, which could worsen the situation. The foundation of a successful emergency response plan and avoidance of catastrophic incidents is based on planning, procedures, training, and communication. An effective emergency response plan should at least have the following features:

- Clear local definitions of what constitutes process emergency conditions to which response must be made.
- Description of the range of responses to emergency conditions.

- Preplanning program based on hypothetical emergency situations. The hypothetical situations should be based on realistic process upsets or excursions that can occur.
- Training and drills, the objective of which are to prepare personnel to deal with emergency conditions in the most effective manner. Emergency conditions place a lot of stress on operating personnel and their ability to think clearly may be impaired. This can be offset by conducting intensive training and drills and by providing refreshers as often as possible.
- The emergency response plan should be regularly reviewed and updated. In addition, major changes to the process or equipment should be followed by a detailed review of the emergency response plan.
- The emergency response plan should provide a description of the hazards of the process materials in a shut down processing system. Any special measures needed (e.g., decontamination, dumping of reactor contents, etc.) should be described in detail and preplanned.

Analyses of recent major accidents have shown that the consequences could have been less severe if the response to the emergency condition had been precise and well-rehearsed. It is vital that in such situations everyone knows where to go and what their specific responsibilities are. The situation is so stressful and tense that the ability to think clearly and logically can be impaired. Since major incidents occur infrequently, every effort should be made to learn appropriate response procedures from incidents and the investigation results of incidents that may have occurred in other plants in the industry or in other industries (e.g., nuclear, aerospace, etc.).

Preparation is the key to successful emergency response. When an emergency occurs, the supervisor can be no more effective than any other person on the operations work team; therefore, the preparation and training of the operations work team for emergencies is crucial and is the responsibility of the supervisor.

7.5. MANAGEMENT OF CHANGE

Process changes are essential in the dynamic process industry. The reasons for such changes may include improving yields, compensating for available equipment, reducing costs, improving safety, and reducing emissions to the environment. A process change can be a change in any of the following:

- Process equipment
- Operating or maintenance procedures
- Operating conditions or process chemicals
- Key personnel involved in the process

The implications for the safety of a process plant of inadequately reviewed changes can be significant, as many accidents have shown. To limit the possibility of accidents, most plants have a formal management of change procedure. Chapter 2 contains a detailed description of the management of change concept and program. The challenge to operations and maintenance supervisors for implementing and practicing management of change procedures is significant. Although most changes are initiated by the operations work team, it is important that all operations personnel understand and recognize changes.

It is essential that first- and second-line supervisors establish and maintain an awareness among the work team that process change should be properly managed to avoid creating unrecognized process hazards. The work team should recognize process change. Once members of the work team can recognize process change, they can follow established management of change procedures and seek appropriate review and authorization before proceeding with the process change. The message that supervisors should emphasize is that "Any change can create a hazard unless proved otherwise." If anything is being done differently from established procedures or outside design parameters, the question should be asked, "Is it a change?" and "Should management of change procedures be initiated?" Thus, the two issues that the operation and maintenance work teams have to deal with regularly are *recognition of change* and the appropriate application of the *management of change procedure*.

An effective management of change program begins with the members of the operations and maintenance work team. Because they are the primary initiators of a change or can recognize a change, it is essential that the supervisor understand the definition of change and why it is necessary to examine any proposed changes. The supervisor should try to convey this understanding to the work team as clearly as possible. In addition, the supervisor should ensure that:

- Process and equipment changes are recognized as they are proposed and that proper review and authorization are sought before the changes are implemented.
- Workers have access to the appropriate persons who can help implement the management of change procedures.
- After the changes are reviewed and authorized, appropriate changes in drawings, operating procedures, training manuals, etc. should be made. Also, any additional training should be completed.

Training workers to recognize changes and to use appropriate management of change procedures can help reduce the occurrence of catastrophic incidents. Worker attitude about this issue and the organizational culture surrounding it are extremely important. Changes should not be viewed flippantly or assumed to not affect process safety. Modifications made to improve plant safety or reduce emissions to the environment should receive the same amount of review and analysis as other changes. The danger is that removing one hazard may introduce another.

Changes do not have to be big and expensive to be process changes that can be dangerous if not recognized and reviewed. Minor changes for example, can be so inexpensive and easily done that they either do not require formal approval or such financial approval can be given at a low level. In effect, such changes may escape formal safety review if someone does not recognize them as process changes. Kletz [3] offers an example shown in Figures 7-13 and 7-14.

Factual description:

A let-down valve was a bottleneck so a second let-down valve was added in parallel. During installation, the check valve was hidden beneath insulation and was not noticed, and the parallel line was joined to the original line downstream of the check valve where there was a convenient branch. The upstream equipment was thus connected directly to the downstream equipment, bypassing the relief valve.

A blockage occurred downstream. The new let-down valve was leaking and the downstream equipment was overpressured and burst.

Contributing causes	Process safety concept
1. When the change was approved, the line diagram was not consulted. Modifications should always be marked on a line diagram before they are approved.	Management of change, Process knowledge and Documentation
2. Those who approved the modification did not check that it was carried out correctly. The person who authorizes such changes should always inspect the finished modification to make sure that the work was done properly.	Management of change, Pre-startup safety review

FIGURE 7-13. Minor modification incident.

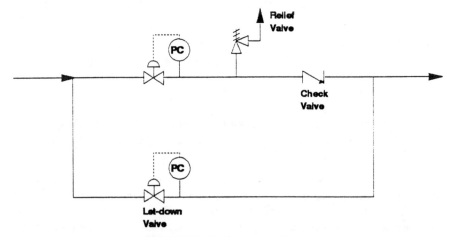

FIGURE 7-14. Modified reactor bypass.

A minor modification could be something as simple as changing an operating procedure step. Many process changes can be easily made; the point is that the persons who can make the change should recognize the process change.

Recent environmental regulations require many changes to processes and equipment to reduce emissions. Although the intentions of these changes are good, they also should be subject to thorough management of change procedures. Sanders [4] summarizes an incident that illustrates this point. The fumes from a benzene tank had to be chemically treated to satisfy environmental regulations. A scrubber was therefore added to the system to treat the fumes. During a plant shutdown, however, the scrubber was taken out of service. Since the steam heating system on the benzene tank was not temperature controlled, it continued to heat up and some benzene vaporized and condensed in the scrubbing system. When the scrubber circulation system was resumed, the liquid benzene and the scrubbing chemical reacted, producing an explosion. The lesson learned is that although pollution prevention regulations required the installation of the scrubber, the change should have gone through a rigorous management of change procedure and process hazards analysis.

Recognizing temporary changes and applying management of change procedures to them is also essential to the process safety of the plant. Supervisors should make sure that even temporary changes undergo the same management of change procedure as permanent changes, and that:

- The time limit for temporary changes is established and strictly enforced.
- Equipment and procedures are returned to their original or design conditions at the end of the time limit for the change.
- Proper documentation of the authorization and implementation of the temporary change is made.
- Appropriate work teams are informed about the temporary change and given all other information associated with the change.

The supervisor needs to remember this key point:

Regardless of how good and thorough the local management of change procedure may be, it can not be put into use unless the persons who are in the position and have capability to make process changes are able to recognize process change and initiate the procedure.

Since persons on operations and maintenance work teams are definitely in position and are capable of making process changes, the supervisor must constantly refresh awareness of process change.

7.6. SAFETY PROTECTIVE SYSTEMS

The purpose of protective systems is to prevent the occurrence of an incident. Some protective systems are intended to mitigate the effects of an incident if prevention systems fail. Protective systems include hardware, software, and administrative systems. A well-trained and disciplined operations work team using current established procedures is an example of an administrative protective system. Table 7-8 offers examples of protective system hardware. Protective systems will vary according to the nature and hazards of a process.

Many safety protective systems are activated by instrument trip-points and it is essential that all trips be kept active. Supervisors should always be alert for and prevent the tendency to disable trips if they have a history to function early, have never functioned in the past, or if the reason for their inclusion in the design is obscure [3]. The supervisor should see that any modification of the trips is formally reviewed and approved according to local management of change procedures. The potential abuse of trips can be controlled to some extent by management procedures. Many plants automatically log all trips and alarms, including the disabling of trips and alarms (safety

TABLE 7-8
Examples of Protective Systems

Protective system	Purpose	Failure scenario	Process safety concept
Process vents	To keep the process under control by venting gas or liquid to the atmosphere or to a holding vessel in another part of the plant.	Disabling nuisance alarms. This action compromises the ability to determine if the system is malfunctioning.	Operating procedures should not allow the disabling of alarms. Management of change procedures should be followed when special circumstances require alarms to be disabled.
Pressure relief devices	Safety valves or bursting devices are intended to operate automatically if some item fails or malfunctions, causing a rise in pressure.	Polymer buildup, or a breakdown in communication regarding isolation.	Operating procedures and mechanical integrity program should include regular inspection and cleaning to avoid polymer buildup.
Explosion vents	In case an explosion occurs, these vents relieve the explosive pressure safely.	Buildup of materials that would prevent its operation.	Operating procedures and mechanical integrity program should include regular inspection and cleaning.
Maintenance drains/vents	These vents are required to prepare for maintenance activities; e.g., vents are necessary to depressure vessels before maintenance begins.	Plugging by scales that form and fall into low points.	Operating procedures and mechanical integrity program should include regular inspection and removal of scales.

interlocks should never be disabled unless specifically allowed in written operating procedures or if management of change procedures have been followed). Such data logs enable an easy historical review of trips that have been made and offer guides for correcting chronic conditions that cause trips or help identify trip systems that are causing nuisance trips. Such data logs offer valuable evidence to investigators if an incident occurs. If automatic logging is not available, trips and alarms should be manually logged. Procedures for temporarily disabling alarms should be formalized in written operating procedures. Some companies use a variety of formal procedures to document the bypassing or disabling of alarms, depending on the safety-critical nature of the alarm.

Operations and maintenance personnel are responsible for ensuring that protective systems remain operational and that operating practices do not subvert the systems. The following sections discuss in more detail two of the most important protective systems: safety shutdown systems and pressure relief systems.

7.6.1. Safety Shutdown Systems

Plants are designed with various emergency shutdown systems that can automatically shut down specific units and isolate parts of the plant or the whole plant, as appropriate. The ability to safely shut down a plant is critical to responding to most emergencies. The primary purpose is to reduce the effect of an emergency on the plant and its surroundings by responding to emergency conditions that can escalate to a catastrophic event.

A system to detect conditions that require an emergency shutdown could consist of an independent detector or sensor, an independent logic switching system, an automatic alarm, or operator observation. For the shutdown system to work effectively, the detection elements must remain functional and independent. In addition, an independent trip system is needed to detect a loss of control that may require shutting down the plant using other manual systems.

Safety shutdown systems are among the last layers of protection against process hazards, upset conditions, and potentially catastrophic incidents. The disabling of safety shutdown systems removes a critical element from the overall plan for ensuring the integrity of a plant's process and equipment. Figure 7-15 illustrates an incident in which the disabling of a safety shutdown system was identified as the major contributing cause. The deterioration of operating practice—pulling the alarm card from the control panel—could have been avoided if process safety principles (i.e., adherence to operating procedures, management of change procedures, and process and equipment integrity) had been followed properly.

The first- and second-line supervisors play a key role in ensuring that the shutdown systems are always operational. If safety shutdown systems must

Factual description:
In a nitrilo triacetate plant, the high-temperature alarm on the reactor was set very close to the actual operating conditions. As a result, very small process fluctuations would set off the alarm and cause a continuous nuisance. The operators gradually got into the habit of going behind the control panel and pulling the alarm card. Once, the cooling water supply malfunctioned, causing the temperature in the reactor to increase. Because an error was made observing the visual temperature indicator, operators thought that the temperature was still within the operating range. However, the temperature had increased enough to cause a pressure surge, which blew the rupture disc. Hot reactants showered on a worker who was seriously burned.

Contributing causes	Process safety concept
1. Process hazards analysis should have identified the setting of the high-temperature alarm as a design flaw and requested engineering analysis and modification.	Process hazards analysis
2. Disabling of the shutdown system by pulling the alarm card.	Process and equipment integrity, Procedures, Training
3. Lack of recognition of a process change and failure to request processchange authorization before disabling the shutdown system.	Management of change, Training
4. Error observing the visual temperature indicator.	Operator awareness, Training

FIGURE 7-15. Incident involving disabling of safety shutdown system.

be disabled, the technical basis for the action and its effects should be carefully verified and the local management of change procedure carried out. The shutdown system should be returned to its normal operating condition when the need for disabling it is over or when authorization for the disabling has expired.

Most plants have action plans for dealing with situations where the safety shutdown systems are out of service. Absolutely key to any such disabling is a preplanned and clearly understood interim protection system that will be in place instead of the shutdown system while it is disabled. Supervisors should make sure, through continuous training and awareness activities, that the members of the operations work team are aware of and knowledgeable about the operational plan and any additional safety precautions required when the safety shutdown systems are out of service. In addition, supervisors should have a clear understanding of instrument loops so that they know which equipment needs to be tested during the shutdown, either individually or as a part of the system.

Some key summary principles for the supervisor to apply when managing the operation work team's use of safety shutdown systems are:

- The team should have a genuine respect for and basic understanding of the facility's safety shutdown systems.

- If a safety shutdown system or the alarms associated with it need to be repeatedly shut down, something is not correct and the supervisor should initiate action to correct the problem.
- If alarms are repeatedly acknowledged and left in alarm, something is not correct and the supervisor should initiate action to correct the problem.

7.6.2. Pressure Relief Equipment

Pressure relief equipment is intended to operate automatically in case an overpressure occurs. The equipment copes with unusual upset conditions to control the incident and avoid a catastrophic release that can occur if the process equipment fails. Pressure relief devices may vent to the atmosphere or to an abatement system, depending on the design and on regulatory requirements. Pressure relief devices, like any other equipment, can fail to operate; however, operations and maintenance personnel can ensure the continued integrity of these systems by continuous monitoring, inspections, and tests, and by avoiding activities that may subvert the system or render it useless. Pressure relief devices are, in many processes, the final layer of protection against containment failure.

The absence of an operational pressure relief system during an upset condition may lead to a catastrophic accident. For example, in one case, a pressurized tank was used to store crude acrylonitrile containing cyanide in an acrylonitrile purification unit. The cyanide tends to polymerize in the vapor phase and can form plugs that can block the relief valve and render it useless. To avoid this, the plant had installed dual relief valves on the tank. In addition, procedures required: (a) the periodic addition of an inhibitor to the tank to inhibit cyanide polymerization, and (b) cleaning of the relief valves every three months. One time the level in the tank was very low and it had not been used for some time. Operators thought that the tank was empty and out of service, so they ceased adding the cyanide polymerization inhibitor and stopped cleaning the relief valves. Over time, the cyanide polymer plugged up the relief valves. In addition, the temperature of the tank contents increased because the polymerization of liquid acrylonitrile caused the pressure to rise. Ultimately, a gasket failed, releasing a cyanide-acrylonitrile mixture that resulted in a large fire. This incident could have been prevented if the relief valves on the tank had not been neglected. The neglect was caused by the assumption that the tank was out of service, simply because it had not been used for a long time.

How can a supervisor avoid a similar incident? Methods that can be applied to the team training process are described in Sections 7.7.2 and 7.8.3. Inspection and testing should be carried out regularly to maintain system integrity. In addition, care should be taken that these systems are never disabled. Close attention to repair procedures and adequate testing can ensure the reliable functioning of pressure relief systems.

7.7. OPERATOR TRAINING

Process operators and other persons directly involved in process operations require basic training in operations skills and in the operating procedures specific to the process. In addition, operators should be refreshed in the operating procedures at appropriate intervals to help them maintain proficiency. The supervisor must understand that no training can ever be a one-shot "vaccination;" periodic "booster shots" are needed.

There is a clear distinction between "training to develop proficiency," discussed earlier in Section 5.3.2 and regular "refreshing of awareness" training in critical process issues. Plants usually have a training program that covers the basic training and refresher training subjects. Management of change procedures and incident investigations should be used as the driving force to determine new training needs. First, management of changes procedures may trigger modified operating procedures or a modified process or equipment, which in turn will require new training. Second, incident investigations may result in recommendations to modify operating procedures, to change or add instrumentation, and to modify processes or equipment, all of which should involve new training requirements.

It is also essential to maintain a clear understanding and knowledge of the process and equipment as the plant experiences personnel changes. One main responsibility of the operations and maintenance departments is to ensure that the work team always has the necessary training. Operators of different areas and units may need to cross-train in certain areas where separate operations are highly interdependent.

The work team changes as individual members of the team change, and valuable operational knowledge can be lost. Accidents can be cyclical simply because operational knowledge has been slowly lost as the work team goes through a "generation of membership." Although there is no cure to a change in membership of the work team, there is definitely a prevention of the unwanted results of loss of operational knowledge. Table 7-9 gives examples of some techniques that can help limit the loss of operational knowledge.

Supervisors usually do not have total control over the makeup of the operations work team. Recognizing this, the supervisor needs to apply the following key principles.

- A cohesive work team is a "process." Change a member of it and the "process" has changed. This process change needs some sort of response to maintain the "process."
- Training and a confirmation of how well the trainees understood the training are the key response.
- Training anchored to an established standard (e.g., established operating procedures) is essential. Training by a fellow team member without that anchoring can create subtle unmanaged "process" change. The supervisor should ensure the anchoring.

TABLE 7-9
Examples of Techniques Used to Prevent Loss of Operational Knowledge

1. Conduct regular refresher training of all staff, including discussions of experiences.

2. Implement a safety awareness program, which should include a discussion of incidents after they occur and on anniversaries of incidents.

3. Plan for significant staff changes carefully. If necessary, the departing staff should be debriefed and the remaining staff should be briefed. Advance planning may also suggest the need for some additional training of the remaining staff.

4. Analyze the operations and maintenance activities of the staff to determine if operating practices follow the operating procedures.

5. Review the operating procedures to determine their relevance and accuracy.

6. Watch for short cuts and "easier" ways of doing things that can quietly slip into daily activities.

7. Review all permit-to-work activities for accurate procedures regarding getting permits and review the work conducted under the permit for the implementation of proper procedures.

8. Consider personal problems (e.g., family problems, health, etc.) when assigning work loads and schedules.

9. Make discreet checks during staff rotation, the departure of experienced staff, and the induction of new staff to verify staff competence and knowledge. Some organizations elect to institute specific testing procedures during these periods.

7.7.1. Refresher Training

Refresher training is the "booster shot" that keeps the original, comprehensive training alive and effective. An effective refresher training program should at least have the following key features.

- Content and frequency of refresher training keyed to understanding the critical issues of the operation.
- Regularity of practice of the refresher training.
- Effectiveness of the refresher training monitored by checking on the trainee's understanding of the critical issues.

Supervisors should make sure that appropriate refresher training is provided when needed. A lack of refresher training leads to a decrease in work team proficiency and a loss of knowledge about critical procedures. Less frequent activities require more refresher training. For example, lack of refresher training about freeze protection procedures for plants located in areas where cold weather is infrequent may lead to hazardous incidents. The following incident illustrates the point.

In a BTX purification unit, water had accumulated in a drain line at the bottom of the column of a distillation tower. During an unusually cold spell that lasted a few days, the water froze. The line expanded and cracked, and when the ambient temperature warmed, a large quantity of benzene was released and ignited. The resulting fire seriously injured two workers.

This incident could have been prevented if timely refresher training about freeze protection had been provided. Refresher training would have required the work team to go through the written procedures, which more than likely would have included a checklist of items to be covered as freeze protection measures. The procedures would also have indicated the need to turn on freeze protection tracing on lines and make sure that exposed lines contained no water.

The content and frequency of refresher training for a particular facility should be decided by persons who know the critical operating issues. Figures 7-16 and 7-17 offer a schematic logic process that can be used to determine the content and frequency of refresher training.

The supervisor may or may not be the person designated in the local management system as the person responsible for refresher training. No matter who is formally designated as responsible, the supervisor has a genuine interest in seeing that refresher training does happen and is effective. The supervisor is the one who needs to see that the key principles of refresher training are applied. The supervisor also has direct control over a fourth key principle: "Make the refresher training fun." The next section offers a way of making it "fun."

7.7.2. Playing "What-If" Games

Training becomes fun when the trainees are genuinely involved and participating. What-if games are an approach that yields effective participation from operations personnel about what can happen in hypothetical scenarios. This informal learning method is based on problem-solving in small groups. The approach allows participants to think through the possible outcomes of varying operator responses to abnormal conditions or of human errors. For example, a cracking furnace what-if scenario could be, "What would happen if steam flow to the furnace was lost?" This scenario would eventually result in the formation of coke in the radiant section of the furnace and possibly a tube blowout, with disastrous consequences.

Other issues that may be investigated during a brainstorming session are: "Would we know if steam flow is lost?" and "What would we do if steam flow is lost?" Given the scenario, the objective of the what-if game is to create alertness about a potential incident and rehearse ways to take corrective measures. For the scenario discussed here, if steam flow cannot be resumed immediately, shutting down the furnace should be considered as an appropriate response.

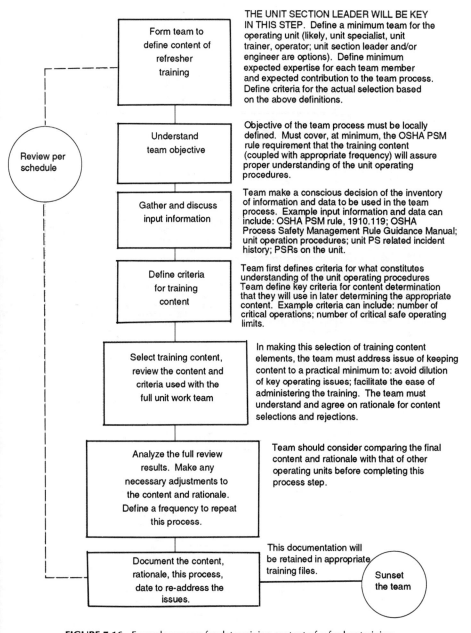

FIGURE 7-16. Example process for determining content of refresher training.

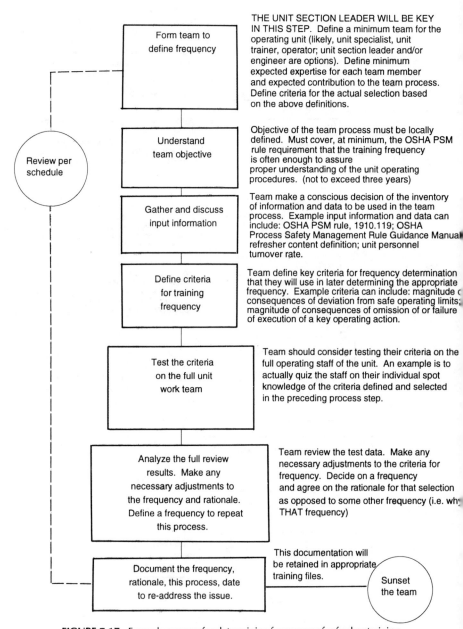

FIGURE 7-17. Example process for determining frequency of refresher training.

The use of what-if scenarios can increase safety awareness and train operators about the correct responses to abnormal conditions. Figure 7-18 is a schematic for the low-pressure stripping of acetylene from an acetylene–ethylene mixture. Process chemistry requires that the temperature at point A remain at or below 215°F. If the temperature goes higher, a decomposition reaction may occur inside the column. The what-if scenario could be, "What-if the steam control valve is stuck open?" A similar scenario could be, "What would happen if the forced circulation pump were disabled?" In either case, the outcome would be that the temperature at point A would start rising until a decomposition reaction started. Therefore, an appropriate operator response to either of the scenarios may be to shut down the process to avoid disastrous consequences.

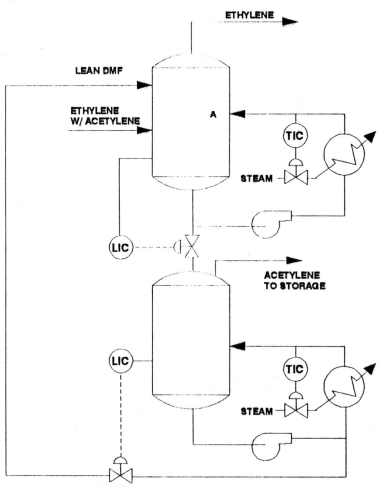

FIGURE 7-18. Low-pressure stripping of acetylene from acetylene–ethylene mixture

7.8. INCIDENT INVESTIGATION

The ultimate purpose of investigating incidents is to prevent the recurrence of the incident. It is far better that a hazardous incident or near-incident never occurs, but if it does a thorough investigation should be conducted. Most plants have established procedures for reporting and investigating incidents. Table 7-10 shows the steps involved in a typical incident investigation and illustrates the roles played by operations and maintenance persons functioning in the investigation process. The first- and second-line supervisors have important roles in the overall incident investigation process. These roles are discussed here.

TABLE 7-10
Incident Investigation Process and the Role Played by
the Operations and Maintenance Departments

Incident investigation steps	Operations and maintenance role
Incident reporting	Most incidents are reported by operations and maintenance personnel. Along with plant management, the first- and second-line supervisors should try to create an environment in which all incidents are reported.
Initiation of investigation	This involves the formation of a team with a clear definition of the scope and objectives of the investigation. Usually, plant or corporate management is responsible for this activity. Among others, operations and maintenance personnel are also chosen as team members.
Collection of evidence	All pertinent evidence is collected. Because of their process experience, operations and maintenance representatives should be able to collect important evidence.
Witness interviews	Besides other witnesses, many operations and maintenance personnel may be interviewed. The operations and maintenance representatives should try their best to produce an environment in which their fellow employees feel free to volunteer information.
Testing and analysis	Operations and maintenance representatives should study the testing results and other analyses to understand and provide input to the process. Examples of testing and analysis are metallurgical tests and consequence calculations, respectively.
Root cause determination	Operations and maintenance representatives should study the results of the root cause determination to determine if the results agree with their operational experience. If it does not, they should voice their opinion immediately.
Investigation report	The operations and maintenance representatives should make sure that the findings of the investigation are complete and presented in a usable format.

7.8.1. Recognizing and Reporting Incidents

Persons who witness incidents need to know the definition of an incident or near-incident so that they can report them and they can be investigated. If an incident is not recognized as such and reported, the opportunity to prevent its recurrence has been lost. The supervisor plays a key role in recognizing and reporting incidents.

Plant procedures will usually have definitions and guides for recognizing incidents. Incidents in which the results are injury, fire, explosion, or a major process release to the environment are not problems to recognize. But a sequence of abnormal events that is somehow stopped short of catastrophic results is also probably an incident that needs investigating. Often, the "stopping short" is simply good fortune or luck. For a hypothetical example, an operator just happens to be in the immediate area to witness a new flange gasket blow out, and he quickly isolates the section of line containing the flange. No injury or damage results. Is the example an incident needing reporting and investigation? Most definitely!

The supervisor should maintain a work team culture in which any abnormal occurrence is considered a potential incident and the decision to report as an incident is a deliberate, conscious, thought-out decision. In effect, if a person decides to not report an abnormal occurrence as an incident, the person or persons making that decision should clearly understand why it is not worth reporting.

It is important that plants establish local criteria for defining and recognizing near-incidents. If local definitions do not exist, the following can be useful to the supervisor when deciding whether to report and initiate an investigation.

- An event out of the expected norm in process operations or process maintenance has occurred. A first response step will be to decide if a catastrophic release, fire, explosion, or injury could have resulted. Examples: unexpected event; unexpected response to an activity. The recognition of the event is an absolute necessity to start the investigation process.
- A process event or chain of events has occurred that clearly could have resulted in a catastrophic release, fire, explosion, or injury but a designed response action successfully prevented the event from resulting in a hazardous accident. An example can be a successful functioning of a safety shutdown system or a pressure relief system in response to a process deviation outside safe operating limits. Some causes of the activation of the protection system may warrant investigation.
- A process event or chain of events was in progress and would clearly have resulted in a catastrophic release, fire, explosion, or injury, but it was successfully stopped. An example can be "almost" opening a misidentified valve. Another example can be an incorrect gasket installed but caught before the process containing the hazardous chemical was started.

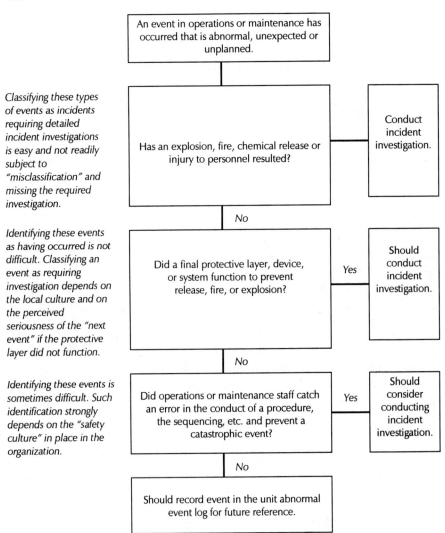

FIGURE 7-19. Incident recognition process generic tool to aid in recognizing process safety incidents.

The important point is for persons who are in positions to witness a process safety incident or near-incident to recognize the event and know how to followup on the recognized events. Figure 7-19 is an incident recognition process model that can be useful. Such a model can capture the local criteria and guide the persons making decisions about investigating the incident. Such a model can be a useful tool to the supervisor when periodically training work team personnel on how to recognize near-incidents.

XYZ COMPANY
INCIDENT REPORT FORM

Date: _____

This form completed by: _____

Signature:_____

Date and time of event: _____

Nature of operations at the time of event: _____

Initiating location of event: _____

Brief description of event sequence: _____

_____

_____

During event, how many people were at the plant: _____

List the names of the people in the plant at the time of the event. Use additional sheets if necessary: _____

_____

Describe status of the following during and after event:

Power: _____

Instrumentation and Control: _____

Other utilities:_____

Equipment: _____

Describe effectiveness of the following during and after the event:

Procedures and training: _____

Emergency response: _____

Emergency shutdown: _____

Safety Manager Action: _____

_____

_____

Signature:_____

Plant Manager Action: _____

_____

_____

Signature:_____

FIGURE 7-20. Example of an incident report form.

Figure 7-20 illustrates an example incident report form. The key to this example and to any form of reporting is to quickly capture the information.

- Who were the witnesses? Who was involved?
- What was witnessed?
- When was it witnessed?
- The known facts about what was witnessed.

Initial reporting of these facts is the first step of evidence gathering and protection of the evidence. The clarity of this initial reporting is important to any subsequent investigation.

7.8.2. The Investigation

Reporting an incident may lead to the conclusion that an investigation is not necessary or that it does not classify as one that could have had serious consequences. Facilities may choose to have simpler procedures for addressing incidents that do not or would not have serious consequences.

Once an investigation has been decided on, plant management may appoint an investigation team charged with formally investigating the incident. The team will usually represent multiple departments so that they can gather the facts of the event, analyze them, and develop plausible scenarios about what happened, and why. The selection of team members is based on their training, knowledge, and ability to contribute to a team effort to fully investigate the incident. Employees in the process area where the incident occurred should either be included on the team or consulted during the investigation, since a close knowledge of events can help the team analyze and determine the underlying causes of the incident.

Operations and maintenance personnel also provide important assistance and support in the following areas of incident investigation.

- *Protection and preservation of the evidence.* After the formal investigation begins, the incident investigation team may develop procedures and a protocol for the long-term protection and preservation of evidence. However, operations and maintenance personnel play a key role immediately after the incident in ensuring that important evidence is not destroyed.

- *Focus on facts, not blame.* It is important to remember that the objective of the investigation is to determine the causes of the incident and reduce the likelihood of a similar incident occurring in the future. Thus, the team should continuously strive to determine the underlying facts and not take the easy alternative, i.e., assign blame.

- *Focus on underlying causes of the incident.* Most incidents can be attributed to human errors and it is quite easy to assign human error as the causative factor and close the investigation. However, all human errors should be analyzed to find the underlying factors. In other words, an effort should be made to answer such questions as: Could the human error have been avoided under different design conditions or operating circumstances? Could work assignments be changed to prevent the recurrence of such human errors?

7.8.3. Investigation Results and Followup

The operations and maintenance representatives on the investigation team should ensure that all pertinent results are clearly summarized in the formal investigation report, which can be used subsequently as a learning tool. The followup to an incident investigation [5] usually consists of the following activities:

- Prompt implementation of any appropriate corrective or preventive measures revealed by the investigation.
- Incorporation of the lessons learned into the permanent operating and maintenance procedures. Changes in the procedures may also require refresher training.
- Communication of the investigation results to all affected employees.

Following are some important points about an organization's "memory" of lessons learned from incidents [5].

- Management should continually relearn and relive previous incidents to protect itself in the present and future.
- It is not lack of knowledge that prevents our safety record from being better than it is. Rather, it is a failure to use the knowledge and information available from past accidents.[6]

The supervisor has a key role in ensuring that the lessons learned from incident investigations are kept in the organization's long-term memory. The objective is to regularly revisit key lessons. One method that a supervisor can use is a working library of operations- and maintenance-related investigation reports that are relevant to the work team. The library should be kept current. The reports in the library should be related to the operations or maintenance activities being practiced and the procedures in use.

The supervisor and others who can knowledgeably participate in selecting relevant reports should first collect local investigation reports. Second, they should review and add lessons learned from available investigation reports from outside the plant and the company. There are many methods by which the investigation reports from the operations unit or maintenance team library can be used to train and to in refresh memory. Two methods are:

- Individually required reading of the reports
- "Tool-box" safety meetings for the work team

Figure 7-21 is a flowchart that offers a training method that has two important advantages over reading and "tool-box" meetings.

- It can satisfy part of the required operator refresher training.
- It can provide an on-going "mini" process hazards analysis of the operations unit or maintenance activity.

A key point of the process is that each person becomes a participant in creating the team product; that is, in developing answers to the questions posed in the process. The questions are:

- Can the incident described in the investigation report happen locally, now?
- If it can't happen now, what is it that prevents it from happening?
- If it can happen now, what ought to be done to prevent it from happening?

An initial team product is a refreshed team understanding of the in-place devices, systems, and procedures that are intended to prevent the incident that was investigated. This is an important understanding. The example training method can produce important conclusions from the team. The conclusions can be reported to the supervisor who assigned the investigation report to the team. These conclusions may be in the form of recommendations for action to prevent occurrence of the incident. In this case, the supervisor will be obligated to: decide if action is warranted; initiate any needed actions; manage the actions to closure; document the closure; and inform the work team of resolution of the actions.

As an example of how the process in Figure 7-21 can be put to use, consider the incident described in Figures 7-7 and 7-8. The conclusions from that incident follow[3].

1. The temperature should be measured in the reactor or as close to it as possible. We should always try to measure the property we wish to know directly, rather than another property from which the property we wish to know can be inferred.

   - The designers assumed that the temperature near the pump would be the same as that in the reactor. It will not be if there is no circulation.
   - The designers assumed that if the pump is energized, then liquid is circulating, but this is not always the case.

2. Operators should not be allowed to change trip settings at will. Different temperatures are needed for different batches. But even so, the adjustment should be made only by someone who is given written permission to do so.

3. More effort might have been made to keep the flow indicator alarm in working order.

4. A high-pressure trip should be installed on the reactor.

5. Operators should be trained to "look before they leap" when they find valves wrongly set.

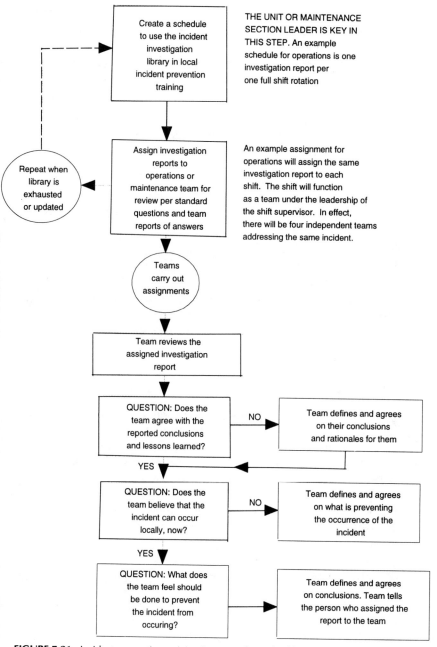

FIGURE 7-21. Incident prevention training (an example method for incident prevention training).

Assume that the work team is involved in operating a batch reaction system. The supervisor can assign the investigation report which will include information like that in Figures 7-7 and 7-8 and the conclusions above to the work team to analyze and answer the "standard" questions. As the work team develops their answers to the questions for their batch system, they learn powerful lessons:

- What currently protects the system.
- What is needed for effective functioning of those protections.
- What might be needed for additional protection.

7.9. HUMAN FACTORS

In well-designed systems where a high degree of hardware redundancy reduces the number of equipment failures, human factors may make up over 90% of the system failure probability [7]. Many studies [8,9] have identified human factors as the overwhelming cause of incidents. Although it is easy to blame human error for an incident, it is a very broad classification and not particularly helpful for developing remedies. In these *Guidelines*, we have therefore made a distinction between human–process interface issues and human behavioral issues.

7.9.1. Human–Process Interfaces

The safe operation of a plant depends on the proper design and functioning of the three-component human–process–equipment system. The interface should be balanced in all of its three parts to ensure the integrity of the process and equipment. In the nuclear industry, it took a major accident such as that at Three Mile Island [10] to focus attention on the need for systematically accounting for the interface of humans with processes and equipment in the control and maintenance of nuclear power plants. Many accidents in the processing industry also can be attributed to a lack of consideration of the human–process–equipment interface. Table 7-11 lists examples of the human–process interfaces that can affect the process safety of the plant.

The objectives of considering the human–process–equipment interface are to enhance the effectiveness and efficiency with which work and other human activities are carried out, and to maintain or enhance certain desirable human values (e.g., health, personal safety). The central approach used in these considerations is the systematic application of relevant information about human abilities, characteristics, behavior, and motivation in executing such functions. In concise terms, any evaluation of the operating procedures and task assignments for monitoring and controlling the process and equipment should be based on their suitability for human beings. The Three Mile Island

TABLE 7-11
Examples of Human–Process Interface Issues That Can Affect Process Safety

1. Are control room workers able to view all necessary information about normal and upset process conditions?

2. Is process safety information available in a format that workers can easily understand?

3. Are worker stations ergonomically well designed so that they can perform their observation and control activities easily?

4. Are related displays and controls of like equipment (e.g., all pumps) grouped together?

5. Do the site layout and control arrangement of similar equipment follow a logical sequence (e.g., Pumps 1, 2, 3, 4 instead of pumps 1, 2, 4, 3)?

6. Are controls easily distinguishable (e.g., distinction between startup switch and shutdown switch; some plants use different color schemes or may require that all shutdown switches be pullout switches instead of push switches)?

7. Is there a system for addressing worker suggestions about changes to the control layout to improve accessibility, visibility, or other factors?

8. Is accessibility to all plant areas taken into account during routine operations and maintenance activities?

9. During the assignment of tasks, are factors such as individual worker capability, and hindrances caused by protective clothing considered?

10. Are different units, pieces of equipment, pipelines, etc. labeled consistently with the operating procedures?

11. Is the system for transferring operational and maintenance knowledge from one shift to another shift adequate?

investigation revealed many contributing causes that could have been avoided by proper consideration of the human–process–equipment interface. Some causative factors of the Three Mile Island incident were reported [10,11] as:

- Incorrect decision to shut down the emergency coolant (human error attributable to lack of proper operating procedures, training, and supervision).
- Incorrect decision related to the status indication of a relief valve (human–process–equipment interface).
- Gross excess of data and information provided to operators (human–process–equipment interface).
- Confusing data format displayed to the operator (human–process–equipment interface).

Computers are a key element in the interface between humans and processes that must be controlled and maintained by humans. Large amounts of information are fed into the computers by various data gathering and data communication devices. The computers manipulate these data and display the converted information to the operators. The operators then use the displayed information to keep track of the process and equipment and initiate any responsive measures that may be necessary. This procedure can become quite complex if the computer malfunctions or fails.

The need to be able to manually monitor and control the process and equipment is very important. Consider, for example, the piloting of commercial aircraft. Nowadays, nearly all aspects of the piloting, including such critical phases as approach and landing, are automated. Speed, direction, vertical speed, and the position of the aircraft relative to the descent axis are all fixed at their nominal values. These automated systems are reliable, fast, precise, and accurate. However, even with the power of the automated system, the pilot cannot relax vigilance of the operation. If the aircraft's automatic controls malfunction or fail, the pilot can take over manual control immediately to avoid a disaster. In the processing industry, computer control may allow operation closer to the operating limits than a manual or single loop pneumatic controller. When the computer fails, operators should evaluate the need to back off on the tolerances and limits under which the process was operating when controlled by the computer. When a computer component fails, operations may require more conservative limits until computer control is restored.

For example, the development of reliable process analyzers and computer systems to quickly process analyzer data has permitted acetic acid to be manufactured more productively from the reaction of butane and oxygen, relying on the oxygen sensor in the reactor outlet to warn of any pending catalyst deactivation. Previous batch analytical methods required operators to shut down the reactor at lower oxygen concentration in the reactor outlet (i.e., operate with wider tolerances on safe oxygen concentration). Failure of the oxygen analyzer or computer control requires operators to resort to the former conservative oxygen limits to ensure safety.

7.9.2. Behavioral Issues

The underlying cause of incidents attributed to human behavior can usually be traced to the compromising of one or more process safety principles. Two types of errors can occur because of human behavior characteristics[12]. These are:

- Errors whose primary causal factors are individual human characteristics unrelated to the work situation. Hiring standards, training, and job assignment policies are ways of reducing such errors.
- Errors whose primary causal factors are related to the work situation. Ergonomic design (typically considered during the design phase), proper working environment, and the efficient scheduling of tasks are some ways that can be used to reduce these errors.

It is possible to reduce the frequency of failure caused by human behavior by applying the principles of process safety. Just as all equipment failures are analyzed carefully to determine the root causes and ways to eliminate them, all failures attributed to human behavior should be analyzed to determine the root causes. The findings should then be incorporated in the process safety management system (e.g., modified operating procedures, improved instrumentation, accountability requirements, modified training programs, etc.) to avoid recurrence of the human-behavior–caused error or to eliminate the effect of such errors on the process. First- and second-line supervisors play a key role in implementing the procedures and practices needed to reduce failures caused by human behavior. In addition, they are the ones who can, by example, influence the behavior of workers and stimulate them to act safely.

Maintaining a high level of vigilance and safety consciousness is always necessary. However, for physiological reasons, the human system passes through a minimum vigilance phrase between 1:00 A.M. and 6:00 A.M. [13]. Some of the more serious incidents (e.g., Three Mile Island, Chernobyl, Bhopal) happened during these hours, which represent a critical period during which the supervisor should ensure that the work team follows procedures.

Many other situations and activities can cause a hazardous incident. Supervisors are in a unique position to detect conditions that may lead to human errors. Because of this, they may also be able to avoid a potentially disastrous situation by taking timely and remedial action. Table 7-12 shows an example checklist for identifying causes of human errors. Table 7-13 gives examples of some key factors that affect human performance and the roles played by supervisors.

When an incident occurs and is investigated, the easiest way out is often to attribute the cause of the incident to a readily apparent operator error, although other factors should be investigated and the root causes should be discovered. A case in point is a disastrous incident that occurred in a Canadian hospital [14]. The incident involved patients who received deadly nitrous oxide instead of oxygen. The problem was traced to a supply line that carried oxygen from a central gas storage system that was inadvertently switched with one that carried nitrous oxide. Although the worker who made the wrong connections obviously made a fatal mistake, an investigation revealed that there were other contributing factors. For example, (a) Why didn't the worker's supervisor identify the pipe mix-up during the installation? (b) Why were the designs of the oxygen and nitrous oxide systems so similar? (c) Why weren't the pipes designed with different connectors? If these questions had been asked and answered earlier and changes made, the disastrous incident could have been avoided.

7.9.3. Spontaneous Response

Frequently, operations and maintenance personnel must respond to situations that require them to take immediate action to maintain the integrity of the process and equipment. These spontaneous reactions can be called "knee-jerk" reactions, but they are dead serious and not comic. When such actions are

---

TABLE 7-12
Example Checklist for Identifying Causes of Human Errors

Human Factors

1. Is the instrumentation and control system appropriate to operators and maintenance technicians? Would computer control ease or magnify the problem?

2. Can an operator isolate a trip system?

3. Is some safety instrumentation unnecessary and therefore confusing?

4. Are all pipes, valves, etc., especially those in hazardous materials service, identified with appropriate symbols?

Training/Procedure

1. What actions are to be taken by the operator upon the failure of any item?

2. What are the consequences if no action is taken upon the failure of any item?

3. Is there a written training policy that applies to all workers?

4. How are retraining needs identified? How are workers trained in new processes, equipment, and procedures?

Behavioral issues

1. For jobs requiring specific physical skills, are workers subjected to pre-employment and periodic health assessments?

2. Are workers health evaluated before they are allowed to return to work after illness or injury?

3. Are there programs to identify and address worker substance abuse?

---

taken, they must be correct. Spontaneous reactions by nature have the potential to increase any hazard associated with the operation. In other words, a hazardous situation causes a response that may correct the problem at hand, but that response may create another problem that could be even more hazardous.

For example, the production of a rubber chemical additive requires that the agitator of the batch system be turned on before reactants are added. This is necessary so that the reactants are consumed immediately in the reaction and the accumulation of excess reactants that could initiate a runaway reaction is avoided. If the operator forgets to turn on the agitator before starting to add the raw materials, a knee-jerk reaction to the later discovery of this operating sequence error can be to belatedly turn on the agitator. This action would probably initiate a runaway reaction or only aggravate the situation if the runaway reaction had already started. Instead, a safer alternative would be to use a reaction quenching agent, dump the contents of the reactor to a safe hold, or use some other mitigation techniques. Even if recovery from the error is complete, the incident should be reported as a near-miss and investigated.

TABLE 7-13
Examples of Key Issues That Can Improve Human Performance

Issues	Role played by supervisors
Motivation	The best motivation can be achieved by making an example. Other motivational techniques are safety meetings, reviewing accidents, commending safe practices, disciplining unsafe behavior.
Work environment	The supervisors should strive to attain an ideal work environment that includes order and cleanliness, minimal absenteeism, and a reduction of work accidents and labor conflicts.
Hiring and management	Evaluate safety attitudes as part of the hiring process. Also, ensure that prospective employees can adapt to the psychological conditions of the company and fellow workers and that they will be receptive to motivation.
Adherence to procedures	Insist on strict adherence to procedures. Every deviation from established procedures should go through a strict authorization procedure.
Psychological state	Consider the psychological state of workers, including stress, family problems, substance abuse, etc.
Training	Training should continually evolve to keep pace with the changing process and new automation techniques. Supervisors should take the lead in providing on-the-job training.
Quality of work crews	When choosing shift crews, it is important to balance the talent of the various crews so that there will be minimal disparity when shifts change. Also, it is desirable to choose compatible crew members to ensure work harmony.

One very useful technique for preparing for knee-jerk reactions is to use the what-if games discussed in Section 7.7.2. The supervisor is responsible for conditioning the operations work team to think before they act, and then to be able to act intelligently and correctly. Constant training and the use of what-if games can help the work team get a more clear understanding of the potential hazards of the work place and of the implications of any responses to those hazards.

Besides normal conditions, operations and maintenance personnel must also deal with a variety of unusual situations, such as abnormal, upset, and emergency conditions to which operators are likely to respond with knee-jerk reactions. For example, a likely response to a shutdown is to restart the process or unit as quickly as possible. Although it is true that delays in restart cause production losses, operators should also remember that if the underlying causes that resulted in the shutdown are not addressed, they may cause a more serious incident later.

Premature restarts can cause a hazardous situation that can escalate to a catastrophic incident. Figure 7-22 illustrates an incident that occurred primar-

---

Factual description:

Figure 7-23 is the schematic of a process to make maleic anhydride from butane, which requires operating in the flammable range for the feed to the reactor bed catalyst. During normal operation, the butane feed pump tripped. The electrician found a bad heater coil in the breaker to the driver of the butane pump. Since the butane level in the vaporizer was sufficient, the unit was kept operational by slowing the reactor to minimum rates, while the electrician replaced the coil. The pump was restarted and the butane level was brought back up and the normal reaction rate was restored. After about one hour, however, the pump tripped again. After several pump restart attempts failed, the unit was shut down. The pump was fixed and unit startup was attempted. When feed was introduced to the reactor, an explosion occurred in the lower level of the reactor. Subsequent investigation determined that the butane–oxygen analyzer and ratio controller were malfunctioning, causing the feed pump shutdown interlock to trip the pump.

Contributing causes	Process safety concept
1. Incomplete analysis of shutdown cause.	Process knowledge, Training
2. Inadequate inspection of the equipment caused the malfunction.	Equipment integrity, Training
3. Procedures should have required verification of the oxygen–butane ratio required for starting up the hot reactor.	Operating procedures
4. Process hazards analysis should have investigated the possibility and consequence of starting with a flammable mixture and recommended the installation of equipment that would have prevented restarting under these circumstances.	Process hazards analysis

FIGURE 7-22. Premature restart incident.

ily because all of the issues were not properly analyzed before operations were restarted after a shutdown. The post-incident investigation revealed that the safety interlock that checked the butane–oxygen ratio was the reason the pump was tripping. In their haste to restart the unit, and because they found a bad coil (the apparent "smoking gun"), the work team failed to check out the whole system. So when the reactor was restarted, the hot reactor and the flammable mixture combined to produce an explosion. Besides checking the whole system, the procedures should have required restarting with a butane–oxygen lean mixture to specifically avoid a flammable mixture.

7.10. AUDITS, INSPECTIONS, COMPLIANCE REVIEWS

Plants should be regularly subjected to safety inspections and audits to detect plant conditions and procedures that could create hazards and to determine compliance with all process safety principles. First- and second-line supervisors, because of their vantage position, can play a key role in conducting inspections and compliance reviews to find areas where the plant's implementation and practice of process safety needs improvement.

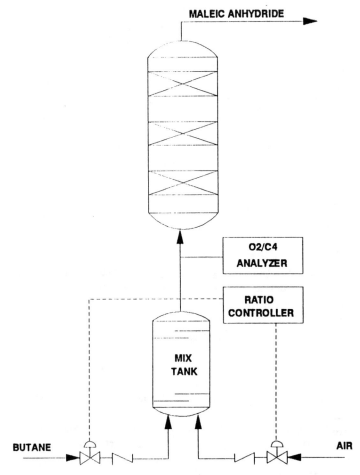

FIGURE 7-23. Schematic of a process to make maleic anhydride from butane.

These audits and inspections are different in nature from the compliance audits required by governmental regulations. Regulatory compliance audits are usually carried out by specialized corporate or plant staff and/or by consultants. In contrast, the audits and inspections carried out by supervisors serve a different purpose (i.e., continual improvement), and represent a continuous effort to improve and perfect the implementation and practice of process safety principles during all activities associated with the operation and maintenance of the plant. The supervisors may conduct these audits formally using written reports to develop an action plan. Alternatively, the supervisor can conduct the audits informally and without creating written reports. Frequent, regular informal audits, along with good participation from the work team with the supervisor and immediate supervisor feedback to the team, are the more effective of the two options of the supervisor's audit.

Some areas that these audits and inspections may cover include:

- *Presence of basic, safety critical equipment and procedures.* These may include sensors, transmitters, controllers, control valves, pressure reducers, etc. The equipment should have current equipment sheets and procedures. The procedures should be available to the people who need them.

- *Regular and effective use of procedures.* The supervisors are in a unique position to informally monitor and audit the use and effectiveness of the operating procedures. First, they should make sure that operators are following established procedures in all of their work assignments. Second, they should try to identify those procedures that need improvement and begin action to change them.

- *Compatibility of operating practices with operating procedures.* Occasionally, drift in operating practices may create a situation where the operating practices may not reflect the operating procedures. Supervisors can prevent this from happening by being the first to detect drift when it begins, and making sure that operating practices are compatible with operating procedures.

- *Compliance with company standards.* Many plant standards are based on company or industry standards. The supervisors, because of their position and experience, should monitor compliance with those standards.

The most important auditing area where first- and second-line supervisors can play a role is ensuring quality operator performance. Procedures and safety programs can be audited by a formal audit program, and inspections and tests can be used to determine the integrity of the process and equipment. However, formal audit programs or inspections are usually not effective for evaluating human performance problems. The supervisors, because of their continuous contact with the work team, can audit human performance issues more effectively. The audits can also be used to ensure that the quality of training is being maintained and that it is suited to the needs and objectives of the operations department. These audits can include detailed analyses of incidents to highlight failures or breakdowns that may have occurred and that can then be used to encourage the maintenance of vigilance and rigor.

Supervisor vigilance and informal audits can increase plant safety and reduce incidents in many circumstances. An incident illustrated in Figure 7-24 is reported by Kletz [5]. During repairs to a pipeline on a pipebridge, for example, workers decided to crawl on planks between the pipes to reach the scaffolding. The incident could have been avoided if the work permit had stated the procedure for conducting the repair and identified the safety

---

Factual description:

During repairs to a pipeline on a pipebridge, workers decided to crawl on planks between the pipes to reach the scaffolding. The pipebridge in question had a walkway on one side, but the pipe to be repaired was on the other side. Scaffolding was erected but access to the scaffolding was a problem. A ladder would have blocked the road, so the two workers were asked to crawl on planks between the pipes to reach the scaffolding. During the repair, there was an unexpected release of carbon monoxide and one worker was overcome. The rescuers had a very difficult job dragging the worker through the pipebridge to the walkway. Fortunately, the worker made a complete recovery.

Contributing causes	Process safety concept
1. Work permit should have outlined the repair procedures and described safety precautions.	Procedures, Process and equipment integrity
2. Deviation from work permit procedures should have triggered management of change procedures.	Management of change
3. A hazard review should have identified the necessity of providing means for easy egress and rescue.	Process hazard analysis, Emergency response plan

---

FIGURE 7-24. Pipebridge repair incident.[5]

precautions. Also, the supervisor should have informally reviewed the work procedures and provided a way for easy egress and rescue.

In its strictest sense, the supervisor's frequent and regular auditing and inspecting is a policing function. It would be unfortunate if the supervisor keeps it as such. What is more important, the supervisor should use regular auditing and inspecting as a training device and as a way to set standards, thus, the supervisor really leads the work team.

7.11. SUMMARY

The main objective of the operations and maintenance departments during the operations phase of the life cycle of a plant is to maintain a safe continuum of operation. To achieve this, the first- and second-line supervisors should ensure that all the elements of process safety are applied appropriately at all times. It is important to remember that accidents usually have more than one contributing factor. These factors can be reduced or eliminated by a process safety program that adds to the multiple layers of protection over and beyond the layers provided by the design and installation of equipment.

A safe plant begins with proper design and installation. Continued safe operation of the plant is achieved by unyielding adherence to process safety principles during all operational activities. The main concept to remember here is that the multiple layers of protection provided by the design, protective systems, operating procedures, training, etc. should not be compromised. As soon as one layer of protection is compromised, the process takes one step closer to an accident. The supervisors, because of their special position, can directly ensure that such circumstances do not occur.

7.12. REFERENCES

1. Cullen, W. D. *The Public Inquiry into the Piper Alpha Disaster*, Volume 1 & 2. Her Majesty's Stationery Office, London, 1990.
2. Rippin, D. W. T. "Batch process planning." *Chemical Engineering*, May 1991.
3. Kletz, T. A. *What Went Wrong? Case Histories of Process Plant Disasters*, Second Edition. Gulf Publishing, 1988.
4. Sanders, R. E. *Management of Change in Chemical Plants*. Butterworth-Heinemann Ltd., Oxford, U.K., 1993.
5. Kletz, T. A. *Critical Aspects of Safety and Loss Prevention*. Butterworths, London, 1990.
6. Kletz, T. A. "Organizations Have No Memory When It Comes to Safety." *Hydrocarbon Processing*, June 1993, pp 88–95.
7. Lorenzo, D. K. *A Manager's Guide to Reducing Human Errors*. Chemical Manufacturers Association, 1990.
8. Rasmussen, B. "Chemical Process Hazard Identification." *Reliability Engineering and System Safety*, vol. 24, Elsevier Science Publishers Ltd., Great Britain, 1989.
9. Butikofer, R. E. *Safety Digest of Lessons Learned*. API Publication 758, American Petroleum Institute, Washington D.C., 1986.
10. Knee, H. E. "Human Factors Engineering; A Key Element of Instrumentation and Control System Design." Proceedings of the 39th International Instrumentation Symposium. Pater No. 93–146, Albuquerque, New Mexico, May 2–6, 1993.
11. Kletz, T. A. "Three Mile Island: Lessons for HPI." *Hydrocarbon Processing*, June 1982, pp. 187–192.
12. Center for Chemical Process Safety. *Guidelines for Preventing Human Error in Process Safety*. New York: AIChE, 1993.
13. Mill, R. C., ed. *Human Factors in Process Operations*. Institution of Chemical Engineers, London, U.K., 1992.
14. Senders, J. W. "Is There a Cure for Human Error?" *Psychology Today*, pp. 52–62, April 1980.

8

MAINTENANCE

The long-term phases of a plant's life cycles are operation and maintenance. Maintenance activities complement operations and contribute to process safety by ensuring the mechanical integrity of the process equipment. Preventive maintenance and predictive maintenance are used by first- and second-line maintenance supervisors to prevent damage and deterioration that used to be accepted as normal wear and tear. Preventive maintenance is inspection and service done according to a fixed schedule. Predictive maintenance is inspection and service based on a schedule that is continuously adjusted according to the real-time conditions of machine performance.

The purpose of this chapter is to increase maintenance supervisors' understanding of the elements of process safety that need to be applied to the maintenance phase of the life cycle of a plant. Maintenance supervisors' roles and responsibilities are defined here and methods are offered for applying process safety techniques. The chapter covers routine and nonroutine maintenance, management of change, aging equipment, critical instrumentation and safety interlocks, training, work permits, management information systems, quality control, contractor safety, and incident investigation.

8.1. ROLES AND RESPONSIBILITIES

Maintenance was once synonymous with repair. Although the maintenance staff had other duties, such as calibrating equipment, their main function was to be on call to make repairs. Preventive and predictive maintenance are now major responsibilities of the maintenance staff. Well planned and executed preventive and predictive maintenance programs will considerably reduce the need for repairs. Together with operations, maintenance personnel determine the equipment, instruments, control systems, and electrical components that are most critical to the safety of the plant process. Ranking the critical items in importance should begin early in the design of the process unit.

The basic responsibility for developing, installing, and administering preventive and predictive maintenance programs rests with the maintenance organization. The execution of approved preventive and predictive maintenance work is generally a responsibility of maintenance personnel and is handled on work orders in the normal planning and scheduling procedures. The operations department has ownership of the plant and is accountable for the safe and economic operations of the facilities. The roles of the first- and second-line maintenance supervisors for preventive and predictive maintenance include:

- Determining the level of maintenance to be applied to the facilities in their custody.
- Ensuring that all equipment is adequately maintained.
- Evaluating results and the performance of equipment to achieve the most economic level of maintenance.
- Coordinating with the operations department to make the equipment available for maintenance service and inspection frequencies.

First- and second-line maintenance supervisors are responsible for executing preventive and predictive maintenance plans that are needed to maintain the mechanical and operating integrity of the plant so that it operates safely and at maximum efficiency.

Operations personnel are responsible for constantly monitoring all systems to make sure that the process equipment is working as designed and to determine when repairs are needed. Operations staff initiate the work orders and make the unit available for repairs; therefore, almost all maintenance repair work should be coordinated with the operations department.

The maintenance department has the primary responsibility for monitoring and controlling the safe performance of maintenance functions. Maintenance personnel are responsible for ensuring that lockout–tagout, purging–inerting, personnel protection equipment, and other safety procedures are adhered to when performing maintenance. The first- and second-line supervisors will usually be the first to recognize the need for additional procedures or changes to existing maintenance practices.

Supervisors have a lead role in upgrading the skill levels of all maintenance personnel. First- and second-line supervisors should evaluate the job knowledge and ability to perform work at each skill level and provide formal and on-the-job training in tasks critical to safe and continuous plant operation. The supervisors are also responsible for refresher training.

A key role and responsibility of maintenance personnel in process safety management must be recognized. The first- and second-line supervisors are in the best position to enforce the complete and proper execution of maintenance procedures. The need to monitor how maintenance procedures are executed is minimal when well trained, knowledgeable, motivated, and alert staff are available, but the need to monitor is always there. Without vigilance, a drift from the complete and proper execution of maintenance procedures can occur and open the door for a process safety incident.

8.2. ROUTINE MAINTENANCE

Routine maintenance is the primary responsibility of the first- and second-line maintenance supervisors. In this section, the performance of routine maintenance tasks and their importance to process safety are discussed, with an emphasis on preventive and predictive maintenance as defined in the introduction.

It is important to distinguish between preventive and predictive maintenance. As valuable a tool as preventive maintenance has proven to be, it is inherently subject to several limitations. In an actual operating environment, variable conditions such as temperature and vibration can accelerate or retard process unit failure. Preventive maintenance schedules do not allow for these variables. Consequently, there is an inherent risk that equipment may be shut down or replaced prematurely, resulting in unnecessary maintenance expense and lost production. Similarly, equipment operated under unusually stressful conditions may fail before scheduled service. When preventive maintenance is not appropriate for a particular piece of equipment, predictive maintenance is often used.

8.2.1. Preventive Maintenance

Preventive maintenance seeks to reduce the frequency and severity of unplanned outages by establishing a fixed schedule of routine inspection and service. Usually these schedules are formulated on the basis of known quantities such as equipment repair history, design life, service plans recommended by the original equipment manufacturer (OEM) and mean time between failures. The chief advantage of a preventive maintenance program is that it gives maintenance management the flexibility to plan and execute required equipment service with a minimum disruption of essential plant operations.

The importance of preventive maintenance to process safety management cannot be overemphasized. When maintenance is performed before breakdown, it can be planned, trained personnel can be available, parts and other materials needed can be available, and many components of emergency shutdown maintenance that result in accidents and injury can be avoided.

Preventive maintenance programs allow first- and second-line supervisors to ensure that the plant and personnel are properly protected when the maintenance is performed. By planning maintenance they can ensure that isolation, lockout/tagout, and other protection is in place before beginning the work. It also allows them to have personnel with the needed training and skill available to perform the work.

Preventive maintenance is the application of systematic attention and analysis to ensure the proper functioning and retard the rate of deterioration of physical facilities. Preventive maintenance activity includes:[1]

- *Operating maintenance:* Properly operating, caring for, cleaning, and in specified cases lubricating equipment. This is usually done while equipment is operating.
- *Shutdown maintenance:* This intermediate segment includes examining, checking, testing, partially dismantling, replacing consumables, lubricating, cleaning, and other work short of overhaul or renovation. Equipment must be shut down to accomplish the inspection or repair.
- *Preventive maintenance overhaul:* Dismantling and examining equipment before a breakdown occurs and replacing or renewing components as they approach a theoretical maximum service limit.

A successful preventive maintenance program assumes the following:

- A set of good maintenance records
- A mutual understanding between the operating and maintenance departments
- Maintenance mechanics who perform in a workmanlike manner
- An adequate inspection program
- A good corrective maintenance program.

First- and second-line maintenance supervisors are in a position to recommend preventive maintenance programs to management. In the past, management has assumed that preventive maintenance programs would result in extra downtime and cost. In practice, the result has not been an increase in downtime or a need to add personnel. For example, Phillips Petroleum Company's natural gasoline department found that preventive maintenance can be justified even if it does not reduce maintenance costs because it does reduce downtime.[2] To establish a preventive maintenance program, both management and the maintenance departments should be committed. Both groups should be convinced that tearing something down when it is obviously running well is justifiable. One way to get both groups committed to a preventive maintenance program is to select some troublesome equipment and show that breakdown can be avoided by preventive methods. Some advantages of a preventive maintenance program are:

- Continuous operation increases
- Reduction of emergencies
- Reduction of expensive labor costs
- Control of parts inventory
- Reduction of production losses

- More satisfied personnel
- Better use of personnel and equipment.

Preventive maintenance should begin during the design phase of the plant or when equipment is selected for replacement or process change. After equipment is installed, inspecting the new equipment to determine if it meets design standards and specifications is the first step in the preventive maintenance program for that equipment.

The first- and second-line maintenance supervisors, because of their knowledge and experience, are often the first to recognize the need to establish a preventive maintenance program for equipment not previously included. They are also the ones who recommend that equipment in a preventive maintenance program be placed in a predictive maintenance program.

8.2.2. Predictive Maintenance

Predictive maintenance, in contrast to preventive maintenance, draws on real-time condition-based data input to continuously adjust the profile of machine performance expectation. Besides temperature and vibration analysis, predictive maintenance systems can use other process variables such as flow rate, revolutions per minute, motor current, and oil analysis. The resulting composite profile is compared to certain program criteria that flag impending service requirements whenever a key parameter is exceeded.

First- and second-line maintenance supervisors may be responsible for gathering data or for verifying data obtained by the operations department for use in the predictive maintenance program. It is also the responsibility of the maintenance department to maintain and calibrate monitoring equipment so that the information obtained is accurate.

Critical equipment and systems should be included in a predictive maintenance program. Complex electronic vibration and thrust sensing analysis is done on large critical compressors because if a rotor is lost the process is going to be down for six weeks. Monitoring is performed to predict failure so that maintenance can be done before the failure occurs. Rotating equipment generally lends itself more to analysis and predictive maintenance than other types of equipment do. Some electronic instrumentation predictive maintenance is based on statistical failure information. When Fisher controls first marketed their compact vertical scale controllers, instrument controllers, and indicator controllers, they provided statistical failure data recorded while the instruments were being developed. The expected failure rates were per thousand hours of operation. The practice within several companies was to replace an instrument if it was in a critical loop and had been operating for a time that approached its predicted failure.

8.2.3. Communication between the Maintenance and Operations Departments

Since most work orders for repair originate in the operations department, almost everything the maintenance department does in a facility has to be coordinated with operations personnel. First- and second-line maintenance supervisors should check that proper communication is occurring between maintenance and operations personnel during the execution of a work order. Operations personnel will usually issue a work order with an assigned priority. Circumstances and conditions dictate the urgency of repairs. Examples of time to respond are: immediate, within twenty-four hours, within a week, within a month, or during the next turnaround. Most chemical operations are organized with one or two shift mechanics, electricians, and instrument electrical technicians assigned to that unit. During the day shift they handle the emergency work orders. If they can not handle all work orders, they get additional help. For instrument problems, instrument people are generally situated close by to ensure immediate service help. For nonemergency jobs, the work order is sent to the maintenance department and assigned to a maintenance planner responsible for that particular area.

The following discussion of work under execution is presented to demonstrate the degree of communication that should take place between maintenance and operations. When the work order arrives, the maintenance supervisor checks with his operations contact, usually the shift foreman. If the unit is large, an operations foreman may be assigned the specific responsibility of planning for maintenance. The maintenance supervisor and operations supervisor will discuss the priority of the job and when it is to be completed. They examine the location to be repaired and the maintenance supervisor plans the job. He lists the parts required, orders the parts, lists the skills and the technicians required, and schedules the work. He alerts the parts people to deliver the parts to the maintenance shop at that time, marked for that particular job. Once the job is planned, the parts ordered, and technicians scheduled by the line maintenance foreman, execution of the job becomes the responsibility of the maintenance line supervisor.

At the scheduled time, the technician picks up the parts and checks in with the operator in the control room and with the operator on the outside, who is responsible for that area where the job is to be performed. The outside operator usually takes the technician to the job area and does a lockout (a detailed discussion on isolation of equipment for maintenance is provided in Section 9.1.1.4). If aware of the repair schedule, the outside operator may lock out the equipment beforehand to isolate and decontaminate it. If hot work is planned inside a vessel or an enclosed space, toxicity, flammability, and oxygen checks should be performed. Before doing internal work within a vessel that normally contains a toxic material, the air within the vessel should be tested specifically for that material to make certain the vessel is fully decontaminated before an entry permit or equivalent is issued. Some facilities designate an independent person from the safety division to perform these tests and to be the final inspector to make certain that all the preparations are correct. The technician then places his lock on the equipment before starting repairs.

Normally, when the maintenance technician is through with the job he removes his lock and notifies the area operator he is finished. The operator or shift foreman usually inspects the completed work before signing the work order. An example is replacing bearings in a pump. The job requires disconnecting the electrical connections and pulling the motor. Coordination is required with the instrument technicians and electricians. Mechanics pull the motor, replace the bearings in the pump and in the coupling box, reinstall the motor, and call the electrician. The electrician reconnects the wiring to the motor and contacts the operator. The operator observes the pump, rotates the pump by hand, replaces the guard, bumps the pump to check for rotation, then signs the work order and removes his lock.

8.2.4. Communication at Shift Change

The continuous operation of manufacturing processes may require multiple shifts of maintenance personnel. When the maintenance personnel change shifts, maintenance-in-progress is literally "handed over" to the next maintenance work team. Ongoing maintenance activities may require that shift change communications also occur between the operations work teams. The outgoing shift needs to communicate necessary information so that the accountability can be properly transferred. The incoming shift should be sure that the necessary information is communicated and understood so that they can properly assume accountability. This needed communication is not likely to be successful if left to chance. First- and second-line supervisors have the key leadership roles during this exchange of information. Also, the supervisors should ensure that other key operations and maintenance personnel exchange information as needed.

Actual methods and procedures for changing shifts are generally established by the local plant. Because of the potential consequences of "missed communication," certain minimum shift change practices are needed that will help ensure consistent process "hand over." Table 7-3 on page 150 in Chapter 7 of this book, shows some critical issues that should be communicated during the shift change.

The Piper Alpha disaster is an example of a catastrophic accident that was largely attributed to lack of communication. On July 6, 1988, a gas explosion occurred on the Piper Alpha Petroleum Production Platform in the U.K. sector of the North Sea[3]. This explosion was immediately followed by a fire in an adjoining production module and subsequently escalated to engulf the entire offshore production platform. A total of 167 people lost their lives in this disaster. The events preceding the initial explosion indicate that a lack of communication between operating personnel and maintenance staff during a shift change was a primary cause of this accident. Figure 8-1 presents a simplified piping and instrument diagram showing the LPG condensate pump system.

Although many other factors contributed to the escalation of the initial explosion into a catastrophe, the miscommunication between operations and maintenance staff was a critical factor leading to the accident. Figure 8-2 describes the incident, referring to Figure 8-1. See Figure 7-3 in Chapter 7 for a description of the Piper Alpha incident seen from the perspective of the operations department.

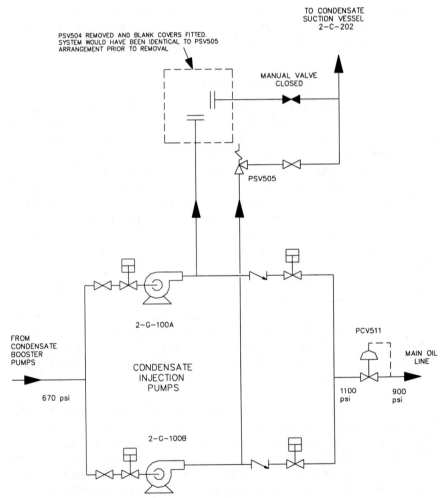

FIGURE 8-1. Condensate pump system on the Piper Alpha platform.

Factual description:

During the day shift, pump A was operating while pump B was on standby. The LPG was flowing through the piping network containing pump A. Operators during the day shift noticed excessive noise emanating from pump A, so they scheduled pump A for maintenance repair services. The operators switched the LPG flow to pump B. A maintenance work order was written to provide service to the now isolated pump A.

Upon review of the maintenance records, a second work order was issued to calibrate the pressure relief valve on the pump A discharge line. During the day shift, pump A service was completed, therefore completing work order number one. However, work order number two on the adjacent relief valve was not completed because a crane was not available to help with reinstallation. The maintenance staff intended to reinstall the pressure relief valve on the discharge of line A the next day.

Shift change on the production platform occurred at 6:00 P.M. The new operators were aware of the maintenance performed on pump A, but did not understand that the pressure relief valve for that piping segment was not completely reinstalled and put back in service.

At about 10:00 P.M. that evening pump B tripped. Efforts to restart pump B failed. The operators then began to switch the LPG flow to pump A but were unable to restart pump A. Electrical disconnects (lockouts) were observed, still in place following the earlier pump A maintenance, so electricians were sent to reconnect the pump motor electrically. Again, the operating staff did not know about the second work order on the pressure relief valve associated with pump A. The power was restored to pump A and the isolating valves were opened and attempts were made to start pump A.

It is postulated that the LPG escaped from the blind flange on the discharge of the LPG pump A during the subsequent attempts to restart condensate pump A. The initial indication of gas from the flammable gas detectors was immediate following the opening of the valves in pump A line. The explosion occurred when the LPG vapors contacted an ignition source.

Contributing causes	Process safety concept
1. The permit-to-work system and shift handovers of work in progress were inadequate. Procedures should have required the departing shift to make sure that the incoming shift knew about the status of all work in progress that would have eliminated the possibility of recommissioning the spare pump.	Accountability, Operating procedures, Process and equipment integrity
2. The diesel fire pumps were on a manual mode, which inhibited the operability of the fire water system.	Equipment integrity, Emergency planning
3. Drills and training for emergencies did not take place with the frequency laid down in the procedures.	Adherence to procedures, Emergency planning

FIGURE 8-2. Piper Alpha incident.

How can maintenance supervisors ensure that safe and successful shift change communication takes place? The supervisor at least needs to understand the local shift change communication procedure and ensure that the maintenance participants in the procedure understand and follow it. This assumes, of course, that a local procedure exists. See Figure 7-4 on page 153 in Chapter 7 for an example of shift change procedures. The number one priority during shift change is for the departing team to communicate the status of any maintenance in progress to the incoming team. The shift supervisors should make sure that this exchange of necessary information is taking place, as required by the procedures and by common sense. Depending on the organizational culture, the shift change communication can be accomplished solely by the individual participants, by means of a group exchange, or by a combination of both. Whatever the method of communication, the incoming shift should clearly understand the status of any maintenance in progress. There should be sufficient overlay in shifts to ensure proper communication with operations personnel during the change.

8.3. NONROUTINE MAINTENANCE

Situations that call for immediate repair are subject to all the problems that occur when work is not planned. Parts may not be available, personnel may not be available, and safety may be compromised. First- and second-line maintenance supervisors should closely monitor work performed under these conditions. It is particularly important that safe-work practices are used and that records are kept. Nonroutine maintenance includes all types of maintenance that is outside preventive and predictive maintenance. The two major categories of nonroutine maintenance are breakdown maintenance and troubleshooting maintenance.

8.3.1. Breakdown Maintenance

Breakdown maintenance is sometimes called fix-it-when-it-breaks maintenance. Equipment that is neither critical to the process nor to the safety of personnel, and that can wait until it fails before it is fixed, is placed in the breakdown maintenance program. When determining what equipment is critical to the process, consideration is given to the availability of spares. For example, the availability of pumps may be necessary to the continued operation of the process. A spare pump is installed next to the primary pump. If the primary pump fails, it is simply a matter of aligning the valves correctly and starting the spare pump. The spare may even be on autostart. The primary pump is therefore one that is only going to be fixed when it is broken. When the seal blows out or starts leaking, it is replaced, or when the bearings start getting rough the bearings are replaced in the box.

Equipment critical to process safety and operability, on the other hand, should be included in preventive and predictive maintenance programs.

Breakdown maintenance of equipment not critical to the process is performed while normal operations continue. First- and second-line maintenance supervisors should be aware of the equipment and operations personnel who are continuing their work in the area. Communicating with operations supervisors during breakdown maintenance is critical.

8.3.2. Troubleshooting Maintenance

Troubleshooting maintenance covers the search for possible hidden causes or multiple causes that can lead to inadequate performance. Trouble shooting looks for any unidentified cause of inadequate performance such as:

- Reduced production capacity
- Inability to meet predicted yields
- Excess consumption of chemicals, utilities, energy, or catalysts
- Unsafe operating conditions
- Operating conditions that require excessive operator attention
- Inability to meet environmental standards.

Since troubleshooting is technically the most difficult aspect of plant maintenance, the most experienced maintenance personnel are used. Supervisors should be particularly alert for hazardous conditions during troubleshooting. It may be necessary to troubleshoot while the plant is operating, making safe-work practices particularly important.

Experience shows that equipment failure, including the mechanical or electrical breakdown of individual process units, is a major cause of trouble. Some trouble-shooting items of interest are:

- Plugging by rust or dirt
- Fallen trays in towers
- Accidental installation of incorrect orifice to measure flow
- Instruments that have not been properly calibrated
- Instrument loop wiring termination errors
- Incorrect assembly of pumps and valves
- Improper anchoring or support of equipment or pipes
- Undiscovered leaks
- Broken agitators
- Freezing because of improper steam tracing or insulation
- Pumps that lose suction because of air leaks in the seals
- Incorrect components in electrical or electronic circuits
- Plugged orifices in burners
- Raw materials and chemicals that deviate from specification
- Faulty design or manufacturing deficiencies.

Examples of faulty design or manufacturing deficiencies include distillation towers that flood when brought to design vapor and liquid flow, equipment that corrodes because an unsuitable material was used in its manufacture, compressors whose motors overload at rated capacity, and clean heat exchangers that display excessive pressure drop.

First- and second-line maintenance supervisors should ensure that records of all failures and breakdowns in the plant are kept. These records are important during troubleshooting because they can help maintenance personnel determine the basic cause of repeated simple failures such as leaks, broken seals, and plugged lines.

8.4. MANAGEMENT OF CHANGE

Inadequately reviewed changes can significantly affect process plant safety. Chapter 2 discusses the mandate for companies to have a management of change program and the elements of the program. It is essential that first- and second-line maintenance supervisors make their staff aware that changes can create unrecognized hazards. Since maintenance personnel are usually the ones making changes to process equipment or initiating changes to maintenance procedures, they are in a position to at least recognize that a potential process change is being made and to be satisfied that the process safety implications of the change have been evaluated.

Maintenance workers should understand what is meant by change. If all maintenance procedures are written and defined, any action outside the boundaries of the written procedure should be considered a possible change requiring the use of management of change procedures. In some instances, maintenance personnel initiate work orders on auxiliary equipment to perform maintenance or to make maintenance easier. Figure 8-3 describes an example of how a change in the location of a tie-in for breathing air nearly caused the drowning of a maintenance worker.

Instrument loops in modern control systems are examples of changes that are easy to make but that can lead to serious incidents. For example, the ranges of transmitters and controllers can be readily changed, but such instrument modifications, despite their apparently minor nature, should be authorized and all concerned should be notified.

Since maintenance personnel actually make most of the changes to the process equipment, they are in a position to recognize the potential for process change and alert management that the work they are performing may require a formal management of change procedure to be implemented. First- and second-line supervisors should train the individuals responsible for plant integrity to be constantly aware of management of change procedures and to recognize if the change they are making should be subjected to these procedures.

8.5. AGING EQUIPMENT

Aging equipment presents a special challenge to first- and second-line maintenance supervisors. Gradual transformations occur as chemical plants age. Combinations of moisture, corrosion, dirt, unauthorized changes, and other unfriendly factors can compromise the integrity of chemical containment and safety protective systems. Therefore, the natural maintenance concern in an

Factual description:

A special system of piping was installed for compressed air to be used with breathing apparatus only. A branch was replaced, but no one appreciated why the original branch was on top of the compressed air main. The replacement branch was installed at the bottom of the supply main. The system was used for years without incident. Then one day a man who was wearing a face mask while working inside a vessel, received a full blast of water that nearly drowned him. Fortunately, he could signal for help and was rescued. The later investigation found that the compressed air main had been renewed and that the branch to the plant had been repositioned at the bottom of the supply main instead of at the top. When a quantity of water entered the system, it drained into the breathing apparatus.

Contributing causes	Process safety concept
1. Management of change procedures were not employed before piping revision.	Management of change
2. Loss of safety memory.	Process knowledge and documentation.

FIGURE 8-3. Breathing air system incident.

aging plant is the increased potential for unplanned releases due to the failure of containment. As plants age, their maintenance should change to combat the effects of aging due to:

- Corrosion
- Erosion
- Fatigue
- Temperature/pressure/flow cycling
- Wear
- Intermittent operation
- Fouling.

All of these factors contribute to the degradation of process equipment integrity. First- and second-line maintenance supervisors should be aware of the results of this aging, because unexpected or undetected weaknesses or failure can have catastrophic consequences. One such instance, described in Figure 8-4, was caused by the failure of an elbow because of erosion. The release of propane at the Shell refinery in Norco, Louisiana resulted in an explosion that killed several people, did millions of dollars damage to the refinery, and affected property several miles from the plant site.

First- and second-line supervisors should develop a management system plan to ensure the integrity of piping and vessels. Normally, this is accomplished by a competent equipment inspection team whose function is to detect any loss in integrity of the chemical containment system. The program should be able to evaluate carbon steel, alloys, specialty metals, plastics, ceramics, elastomers, and other materials before conditions deteriorate to the point where there is a loss of chemical containment.

Factual description:
In the depropanizer overhead line at the Shell Norco Louisiana Refinery, an ammonia–water mixture was injected prior to the condensers to control corrosion. A debottlenecking project added another set of condensers, in parallel with the original units. Another ammonia–water injection system was added to the new piping upstream of the new condensers. The injected ammoniated water was directed too close to an elbow and the eroded pipe suddenly failed several years later. This piping failure in a large hot pressurized distillation tower overhead line caused widespread damage to the entire catalytic cracking process area.

Contributing causes	Process safety concept
1. No pre-startup safety review was conducted to ensure the proper placement of the corrosion inhibitor injection port.	Pre-startup safety review
2. Inadequate inspection and interpretation of pipe wall thickness measurements did not reveal the eroding pipe.	Mechanical integrity, Equipment inspection

FIGURE 8-4. Norco incident.

Table 8-1 illustrates monitoring techniques that can be used to determine the condition of plant and equipment. The monitoring should include the safety protective systems as well as equipment. Monitoring techniques may combine physical measurement with regular and frequent patrols. First- and second-line maintenance supervisors should be certain that the monitoring systems, as well as the equipment, are maintained as the plant ages.

TABLE 8-1
Monitoring Techniques for Determining
Condition of Plant and Equipment

1. Vibration monitoring
2. Thermography
3. Ultrasonic thickness testing
4. Measurement of heat transfer performance
5. Process unit performance monitoring
6. Control valve position measurement
7. Scheduled testing of safety protective system
8. Acoustic monitoring
9. Lube oil analysis
10. Corrosion coupon
11. Interstage temperature monitoring of compressors
12. Flue gas analysis of fired heaters
13. Strain gauge measurements
14. Bearing temperature trend
15. Electrical usage trends

8.5.1. Corrosion, Erosion, and Fatigue

Corrosion, erosion, and fatigue can result in failures resulting in process gas or fluids entering unexpected places or being released. Because such a loss of containment of process gas or fluids can endanger the plant and personnel, plant piping and vessels subjected to harsh chemical attack are the ones that should receive the most attention from the maintenance staff over the life of the plant.

Corrosive systems, such as hot brine, sulfides, and chlorides, should be monitored. The corrosion of secondary pressure containment, such as steam-jacketed pipelines or canister-type motors, may allow process fluids to escape. Temporary piping not designed for corrosive environments may degenerate rapidly, increasing the risk of flammable or corrosive fluid leaks that can weaken the integrity of other equipment or even structures. For piping and vessels handling noncorrosive fluids, cathodic protection at effective levels requires diligent watchfulness by maintenance personnel to ensure that equipment such as rectifiers continues operating and that other sources continue to produce adequate amounts of current. Regular checks should be made and inspection records should be kept.

Maintenance personnel should be aware that corrosion under insulation can cause troublesome pitting or the lamellar rusting of carbon steels and chloride stress corrosion cracking of austenitic stainless steels. This hidden metal loss may result in drips, spills, ruptures, and environmental releases that can allow flammable material to escape and cause fires and explosions. Brackets that support insulated vessels can collect water. Studies[4] determined that the most severe corrosion problems occur under thermal insulation at temperatures between 140–250°F, that is, too low to quickly boil off any intruding water. At higher temperatures, other corrosion problems can occur. The intruding water can carry chlorides and other corrosive elements that can concentrate and result in stress corrosion cracking.

Erosion can occur from liquid flow and is one of the most difficult types of damage for maintenance personnel to detect. If the staff know where it is likely to occur, nondestructive thickness testing can be used to test the material. Injection ports are particularly suspect, since a liquid sprayed into a flowing gas stream may cause erosion. Figure 8-5 shows a corrosion inhibitor injection port that was placed too close to a piping elbow. After several years in service, the elbow catastrophically failed because of the loss of metal caused by liquid impinging on the pipe elbow.

Even small droplets of liquid can damage piping when these droplets are accelerated to high velocity. They can erode pipe if they are particularly abrasive. The special name for steam droplet erosion of valves is "wire drawing" because the erosion looks like a wire drawn across the valve seat.

A common example of fatigue failure involves heavy valves on small-diameter pipes. Although fatigue is generally a slow mechanism, the final

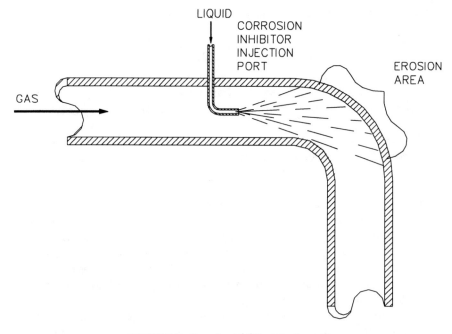

FIGURE 8-5. Corrosion inhibitor injection port.

failure can be sudden. Pressure, temperature, or flow cycling can lead to the growth of fatigue cracks in pressure vessels and piping systems.[5,6] Hence, it is necessary for first- and second-line supervisors to ensure that the equipment is closely monitored for those chemical processes that cycle.

Water hammer occurs when a slug of fluid has to negotiate a sharp turn or is stopped by the sudden closure of a valve. This water hammer phenomenon can loosen pipes and joints and even break pipe. Maintenance personnel should pay extra attention to any equipment that has been subjected to water hammer. Heat exchangers are especially vulnerable to water hammer because some units have relatively thin walls.

8.5.2. Wear, Intermittent Operation, and Fouling

Wear is usually associated with moving or rotating machinery such as pumps and valves. As the impeller in a pump wears, the flow characteristics change. Wear in a valve may eventually cause a leak that would lead to inadequate chemical isolation. Figure 8-6 describes an incident of a worn valve that led to an explosion in a reactor that killed two workers.

A process unit subjected to intermittent operation, such as a batch process, may have a shorter life expectancy because of the stress of the variable

Factual description:	
The reactors had been properly cleaned and the gas freed before a maintenance operation was initiated within the vessel. Since there was no plan to use hot work in the vessel, the plant supervisors decided not to use blinds for isolation but relied simply on closed valves. Some flammable vapor did leak into the vessel, since the valve was sufficiently worn to allow leakage, and this flammable vapor leak was ignited by a grinder.	
Contributing causes	**Process safety concept**
1. Failure to follow safe-work practices, especially checking the vessel for the presence of flammable vapor.	Hazard identification, Maintenance Procedures
2. Failure to recognize the grinder as a possible ignition source, thus requiring a hot-work permit.	Hazard identification, Standards and Codes

FIGURE 8-6. Worn valve incident.

operating conditions. Instruments and electronic controls are more susceptible to damage from transients, which occur with repeated startup and shutdown actions. Therefore, this equipment may age more rapidly and should be placed on a more frequent maintenance schedule.

Piping and heat exchanger fouling is usually a gradual mechanism that should be monitored by maintenance staff over the operational life of the process. Biological fouling of the cooling water, process water, and fire water systems is common, particularly if plant biocide procedures are interrupted. Reduced performance caused by the fouling of a fire water supply system can have a direct effect on process safety. If the reduced water pressure or reduced water flow rate causes inadequate equipment cooling during an emergency, the potential for escalating a small incident into a major accident exists.

8.6. CRITICAL INSTRUMENTATION AND SAFETY INTERLOCKS

First- and second-line supervisors should set the priorities for maintaining resources. Any work requests for critical instruments or safety interlocks should receive high priority since they are crucial to plant safety. Many chemical process plants depend on "engineered safety" and have numerous interlocks, relief valves, and other protective devices to safeguard against different hazards or accident mechanisms. The chemical processing system is, therefore, safe as long as the protective devices operate on demand; however, it is possible that the devices fail on demand, allowing an accident to occur. Protective devices can fail for many reasons, including wear and tear, corrosion, human error, lack of capacity, and environmental conditions.

Traditionally, chemical processes have relied on instrumentation to operate safely and economically. From the age-old tank dip stick to a modern high-level alarm, the purpose of some instrumentation has been to give early warning of potentially unsafe conditions. The same instrumentation may serve both operator control and safety functions. Instrumentation systems that protect against unplanned chemical releases or that help maintain chemical containment integrity have been categorized as "critical" instrumentation. A high-level alarm on a flammable liquid storage tank may be deemed critical, whereas the same type of high-level alarm on a water tank may not be.

In more hazardous or more complex chemical processes, separate instrumentation systems have been developed to enhance the process safety function. Figure 8-7 shows the conceptual distinction between process control instrumentation and special safety instrumentation, called safety interlocks. An emergency shutdown system, the safety interlock is a device that prevents process equipment from being placed in an undesirable or hazardous condition. Safety interlocks are designed to protect process equipment from exceeding safety boundaries, thus keeping the chemicals contained. Safety interlocks may either divert chemicals to an alternate system or shut down the process equipment to within the safety margins. Some companies further distinguish between safety interlock systems that maintain chemical containment (such as high-temperature or high-pressure shutdown systems) and emergency response systems (such as fire and gas hazard mitigation systems) activated after the chemical containment is breached.

Safety interlocks may be electrical, electronic, pneumatic, hydraulic, or even accomplished by computer software. A few examples include:

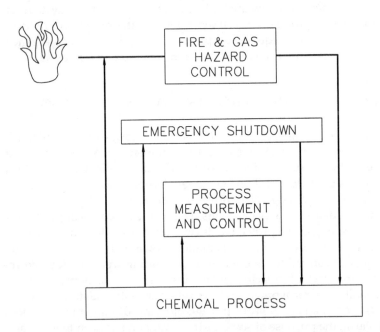

FIGURE 8-7. Process instrumentation and special safety instrumentation.

- During the startup of process machinery, interlocks are used to ensure that all prestart conditions (adequate lubricating oil pressure) are correct and that the correct sequence is followed.
- Interlocks are used to prevent unauthorized entry to special areas, such as electrical switch rooms and process vessels.
- Isolating valves fitted with pressure relief valves may have interlocks installed to prevent them from being closed simultaneously.
- Micro switches and solenoid valves are widely used to actuate alarms and shut down mechanisms; they are common components of electro-pneumatic interlocks.

Because of the important nature of critical instrumentation and safety interlocks, first- and second-line supervisors should be particularly alert to when maintenance is needed on this equipment. The DuPont[7] system allows a range of authorization levels for the maintenance of safety interlocks. For example, a process foreman may bypass a noncritical instrument loop for maintenance, such as a lubrication alarm, but other, more critical alarms may not be bypassed by anyone.

Today, many chemical processes are controlled by programmable electronic systems, digital control systems (DCS), supervisory control and data acquisition (SCADA), and programmable logic controllers (PLC). Programmable electronic systems include the single monolithic computer installed in the control rooms of the 1960s and 1970s, as well as the modern network systems of electronic devices linked on data highways throughout the chemical plant. These modern programmable electronic control systems, as depicted in Figure 8-8, contain many components that work to provide the safety interlock function. The potential for safety interlock failure increases with increasing system complexity.

In a highly automated plant, it is surprisingly difficult to pinpoint the source of a malfunction and distinguish between cause and effect. Many misconceptions are possible. The Three Mile Island[8,9] nuclear plant incident was exaggerated because plant operating staff did not recognize the implications of a pressure relief valve position alarm, which was not directly measuring the valve position but the status of the solenoid that powered the valve. The level of training required to maintain these complex computer-based safety interlocks requires that most chemical process facilities designate certain maintenance staff with special training to be responsible for these systems.

8.6.1. Proof Testing

Safety interlocks are tested periodically, typically every 6–12 months, according to specific test procedures. The two methods of testing safety interlocks are an actual test and a simulated test. The actual test is performed by operating the process to the trip point, ensuring that all instrument compo-

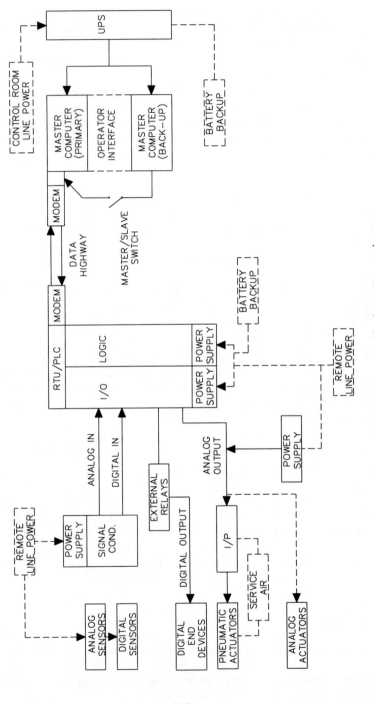

FIGURE 8-8. Programmable electronic control system.

nents function properly. The simulated test is used when the safety interlock instruments are so numerous and complex that a single trip shutdown cannot verify all aspects of the safety systems. The simulated test, however, cannot substantiate the safety interlock capability as thoroughly as an actual test. Testing an interlock system consists of more than simply verifying that the interlock works. A functional test really inspects the whole system from sensor to control actuator for potential deterioration. Maintenance and engineering staff should coordinate when developing test procedures for complex interlocks.

An appropriate authorization system, which enables critical instrumentation to be maintained and calibrated, is an important part of the proof testing protocol. The test frequency for each critical instrument system should be determined from engineering and maintenance studies. Certain safety instrumentation components, such as emergency shutdown valves and process isolation valves, require the same maintenance and lubrication as other, similar mechanical devices in the chemical process. Since these devices are not normally in operation every day, they may contain hidden faults that the operators do not recognize. In some organizations, maintenance personnel are responsible for determining the operational capability of the safety systems.

When a safety interlock has failed silently to an unsafe condition such that it is unable to respond to a process demand, such a failure is detectable *only* by either: (a) failure of the system to respond to a real process demand, or (b) proof-testing before a real process demand occurs. For example, if proof-testing is done on an annual basis, the system can be in a failed condition with the plant unknowingly at risk for a long period. Proof-testing frequency should therefore be established by consideration of (a) possible system failure frequency, (b) possible frequency of process demands on the system, and (c) the criticality of the system (i.e., the severity of consequences of failure to respond to a process demand). Test frequency must be substantially greater than system failure frequency; otherwise, the system is *likely* to fail between tests. Similarly, test frequency must be substantially greater than the frequency of process demands on the system if testing is to serve a useful purpose; otherwise, the process will do the testing—with adverse results. Lastly, the more critical the instrument, the more important testing frequency becomes. Process hazard analysis specialists—and, ideally, system designers—can use numerical analysis to establish proof-testing intervals that will provide an acceptable level of risk (*or hazard rate*).

8.6.2. Critical Instrumentation and Interlock Classification

Once a plant has decided to categorize the critical instrumentation and safety interlocks and the criteria have been established, maintenance departments usually develop prioritized lists of critical instruments and safety interlocks. The criteria used for the classification should be documented in the plant maintenance records.

The classification systems used for critical instruments and safety interlocks differ from company to company. For example, at their Orange, Texas plant, DuPont[7] has divided their critical instruments into two groups: operator aids and safety interlocks. The operator aid is used to alert the operator to an abnormal, but generally nonhazardous condition. The safety interlock alerts operating staff to potentially hazardous conditions or maintains the process in a predetermined stable state.

On the other hand, Monsanto[10] has classified their critical instrumentation and safety interlocks into four separate classifications:

- Community protection safety loops
- Employee protection safety loops
- Major property and production economic loops
- Minor property and production economic loops.

Monsanto statistics indicate that eighty percent of all safety loop failures are due to failures of the field devices, and these are roughly evenly split between the sensor and the actuator.

8.7. MAINTENANCE TRAINING

All maintenance personnel working in a chemical processing facility should be trained in an overview of the process that gives them a basic understanding of the hazards associated with the process. There are many other types of training, including mechanical skills training, theory training, on-the-job or apprenticeship training, safe-work-practices training, and specialized equipment training. Maintenance staff who are expected to perform unsupervised maintenance jobs on process equipment should be trained in written maintenance procedures and show competency in performing the required tasks. Demonstrating competency may be achieved by written tests, oral exams, or observed demonstration.

Training is the responsibility of first- and second-line supervisors. To be effective instructors, line supervisors should know their subject and be competent in the use of effective instructional techniques. To become more effective, instructors should take courses in instructional techniques that cover the following:

- Instructor attitudes
- Student similarities
- Motivating in training
- Analysis of material
- Use of visual aides
- Active participation
- Informal evaluation
- Use of progress tests.

During the course, each participant should be given opportunities to lead practice training sessions, thereby obtaining a thorough understanding of the meaning and importance of the principles that apply to practical training sessions. Each participant should recognize that he or she can effectively discharge this often unstated responsibility for training.

Two main factors contributing to the success of maintenance personnel training are the organization of the instructional material and the quality of presentation, both of which are critical to effective and efficient training. The instructional material will cover such a diverse array of maintenance skills that it is best organized in lesson units, each unit covering a particular topic. The topics should cover a wide range, from the installation of roller bearings to precision surface grinding practices.

Each topic should be independent and complete in itself. Specific training goals should be stated and organized in good teaching order. Each topic should be designed for an instructional period of not more than two hours. The material contained in the lessons may be prepared by personnel experienced in engineering, plant maintenance, and in industrial training. The objective is to put the maintenance information in practical form for everyday use in the hands of line supervisors. If the training material is not available in practical form, it should be developed so that training activities can be scheduled with little advance planning. With this type of material, each unit and group within a large plant maintenance organization can be very selective about arranging material to fit their specific maintenance training needs. In the same way, instructional material can be readily arranged to fit each level in the plan of progression from helper to higher positions.

Job knowledge can be measured by written tests. Ability to perform is the purpose of job knowledge. The candidate for a maintenance job position can demonstrate this ability by performing an actual plant job arranged as a check task. The task should be carefully defined and the individual should be informed about what he will be evaluated on. The task should be given to him as if he were being assigned a job in the plant. These qualification checks, both of job knowledge and skill performance, provide an objective measure of an individual's qualifications for a maintenance job position.

8.7.1. Upgrade and Refresher Training

The systematic upgrading of maintenance skills is an essential ingredient of an effective maintenance program. While in the position of helper, and at each subsequent level, the mechanic should receive both formal and on-the-job instruction in the maintenance tasks pertinent to the safe and continuous operation of the plant. Before advancing to a higher level, each mechanic should qualify himself under a systematic procedure by demonstrating job knowledge and an ability to perform the work. An upgraded training program should be initiated before the startup of a new facility to help develop the full qualifications of new employees. New applicants that are hired for a position above entry level should demonstrate that they possess the required skill, knowledge, and performance ability. Applicants can then be placed at a level consistent with their experience and abilities and consistent with the requirement of the maintenance tasks for the plant. Refresher training for maintenance personnel should emphasize safety. Safety training should cover safe-work practices, personal protective equipment, maintenance procedures, lockout/tagout, work-permit protocol, confined space entry, management of change systems, and emergency response.

8.7.2. Loss of Plant-Specific Maintenance Knowledge

Maintenance staff often change after a period, taking with them valuable plant maintenance knowledge. Similarly, staff may be promoted or leave the company, diluting the pool of knowledge. Several accidents have occurred because of "corporate memory fade."[11] Even when staff does not change, over time they may drift away from strict adherence to procedures, or become careless. Although no universal rules exist for solving these problems, each organization must find its own solution within the limits of its resources. The following techniques, for example, have proved helpful in limiting memory loss and fade:

- Refresher training of all staff
- Effective use of incident investigation in training
- Publicity campaigns reminding staff of accidents on the anniversaries of the accidents
- Training and supervising new staff until they can demonstrate competence
- Reviewing maintenance instructions to evaluate their relevance and accuracy
- Walking the plant and observing if short cuts are slipping into everyday use
- Talking to the mechanic and asking him why he is doing a job "that way"
- Checking work permits for accuracy and then visiting the workplace to see if the work is being done as required
- Being aware of staff who have problems off the work site, such as a family crisis in progress; performance may be so impaired as to be hazardous

8.8. WORK PERMITS

For all types of maintenance, the safe completion of the job depends on proper preparation. The work permit approach to preparing a plant for maintenance allows the hazards involved in a job to be properly determined. Work should never be carried out in a potentially hazardous area without safe-work procedures. These procedures should ensure that persons doing the work and others associated with them are not exposed to danger, that the work does not cause danger to others not directly concerned with it, and that everyone is complying with regulations, technical standards, and codes of safe practice.

The following principles form a basis for a work permit system:

- Complete and secure isolation of equipment
- Complete removal of any residual hazard
- Clear and correct identification of all equipment in the plant
- Training and instruction of maintenance workers
- Monitoring of procedures
- Authorization of changes in procedure, materials, tools, and personnel.

First- and second-line maintenance supervisors are responsible for seeing that all maintenance work, however minor, is properly authorized, controlled, and recorded. When sections of a plant are being isolated, up-to-date piping and instrument diagrams (P&IDs) should be available before the work begins (see Section 8.9.3). A work permit procedure is an acceptable way to meet these requirements; however, the system can become too complex, and if it is used unnecessarily, its value will diminish. To avoid this, work that is precisely defined and clearly not covered by the work permit system is better carried out under the control of its own system of documentation.

All maintenance personnel should be familiar with the work permit system. They should recognize when maintenance work requires a work permit before it is performed. First- and second-line maintenance supervisors may be responsible for verifying that the work permit is needed and that it is properly executed. Company policy and maintenance procedures determine the types of maintenance that require a work permit. For example:

- Confined spaces
- Where toxic, flammable, or corrosive substances are present
- Where a lack of oxygen or oxygen enrichment can occur
- Where hot-work is to be done on plant or equipment
- Where lines or vessels are to be opened
- Where excavation, such as trenching, is to take place
- Locations where the accidental or unauthorized starting of the plant or equipment may endanger others
- Conveyors, lifts, hoists, and cranes.

The preparation of a work permit involves two groups of people: maintenance staff and operating staff. Personnel responsible for issuing work permits may be drawn from both these groups and should be appointed by virtue of their training, their experience at the plant, and other relevant qualifications, rather than because of any position held by them in the organization.

To safely complete a job, the maintenance staff should be well prepared to ensure that no important safety matters are overlooked. Preparations for maintenance work should be detailed in procedural checklists. If each step is itemized, important steps will not be forgotten or missed during shift changes. Figure 8-9 describes an example of a simple pipe repair that resulted in an accident because of sloppy work permit procedures.[12]

For a work permit to meet maintenance needs, it should include the following identifying information:

- The name and address of the company to which it belongs and, if different, a reference to the premises where it will be used (the form fulfills a legal requirement and may be required as evidence)
- An individual and exclusive reference number
- The date when the work may be done and the starting and finishing times
- Positive and unambiguous identification of the location of the job and the equipment to be worked on
- Clear details of the work, the means of work, and the number of persons covered by the permit.

The work permit form is used to document the preparation and completion of maintenance work. Table 8-2 illustrates important safety features of the work permit form; however, no one design is suitable for all companies or all purposes.

It is common practice in the chemical industry to pay special attention to hazardous types of work, such as confined space entry or the use of open flames (hot work). This is done by using forms, often of different colors, dedicated to these activities. For example, a welding operation inside a vessel would require a general permit to cover vessel cleaning and isolation, as well

Factual description:	
A fitter was to repair a leaking joint in some pipe work carrying water on a pipe bridge. Staging was erected, but because of the difficulty of access, the process supervisor pointed out the joint to the maintenance supervisor from the ground. The maintenance supervisor, in turn, pointed it out to the fitter. The fitter opened a joint in a carbon monoxide line in error, was gassed, and fell to his death from the staging.	
Contributing causes	**Process safety concept**
1. The work permit should have outlined the repair procedures and described safety precautions.	Procedures, Process and equipment integrity
2. Deviation from work permit procedures should have triggered management of change procedures.	Management of change
3. The joint to be repaired should have been tagged for identification.	Process hazard analysis, Communication

FIGURE 8-9. Pipe repair incident.

TABLE 8-2
Important Safety Features of a Work Permit

Features	Explanation	Reason/comment
1. Description of work	What has to be done, why, and how it is to be achieved	Simple statement of the work and the method to be used
2. Safety requirements	Those precautions that ensure the safety of persons, plant, and product	Should be listed specifically; avoid general statements
3. Recognition of hazard level	This is a positive way to make the initiator aware of high-risk situations	Relates to both place of work and nature of work
4. Acknowledgment of high-risk hazard	High-risk situations often require and warrant more than a single viewpoint	Consideration by and input from other disciplines warranted
5. Indication of work progress	The commencement of work may reveal a far more complex job than originally envisaged	Provides an opportunity for job closedown and reappraisal
6. Authority to start	The signature of a person properly qualified to begin work	Persons should be specifically identified; not by position alone
7. Acceptance of conditions	The signature of the person responsible for the work undertaken	This person is then responsible for others
8. Acknowledgment of completion	The signature of the person responsible for 6 above	Restoration of normal functions

as a hot-work permit. Because of the special risks involved in such operations, fewer people should be authorized to sign entry and hot-work permits than general permits.

These high-risk operations may require a member of senior management to verify both the work content and the safety precautions. This verification may include a meeting of all involved parties before the work starts. Attendees might include process operators, maintenance staff, and laboratory technicians. If this procedure is carried out, it should be recorded on the permit.

The issuer of the permit is the one who has the authority to authorize the work. Acceptance of the authorization with its regulating conditions is the responsibility of the person requesting the permit. Both the issuer and the requestor should sign the form. Similarly, the completion or termination of the work should be formally acknowledged by the issuer, who should then sign again.

ACME CHEMICALS HOT-WORK, ENTRY PERMIT

Serial _____ Page 1/3

To be used for entry into a confined space and/or work
involving a source of ignition.

Contractor _____
Pass No.   _____
Site contractor _____
Tele. No.   _____

---

1. LOCATION _____

_____

Area classification _____

2. PERMIT VALID FROM TIME OF
CERTIFICATION UNTIL

Time _____   Date _____

---

3. WORK TO BE CARRIED OUT

_____

_____

_____

4a. ISOLATION—Details of isolations carried out	Time	Date	Initial
_____			
As per Plant Maintenance Preparation Procedure No. _____ Other isolation/preparations carried out on cold work permit No. _____	_____	_____	_____
	_____	_____	_____
4b. ELECTRICAL ISOLATION—Details of isolation carried out _____ _____	_____	_____	_____
	_____	_____	_____

---

5. ATMOSPHERIC TEST CERTIFICATE—I certify that the work area CAN/CANNOT be entered without
breathing apparatus and IS/IS NOT safe for means of ignition as specified in section 6 below.
NO GAS TEST DONE/GAS CERTIFICATE ATTACHED
Certificate No. _____   Signed _____ Time _____ Date _____

---

6. OPEN FLAME CERTIFICATE Type permitted Open flames or other ignition _____ sources ARE/ARE NOT permitted _____	Fire protection required _____ _____

7a. HAZARDS REMAINING _____

_____

_____

7b. PRECAUTIONS TO BE TAKEN (Continue on back if necessary) _____

_____

_____

_____

_____

---

8. CERTIFICATION—I certify that the precautions in sections 4 and 5 have been taken and that work may be
subject to the requirements of sections 6 and 7.

Signed _____   Position _____ Time _____ Date ____

9. ACCEPTANCE (Continue on back if necessary)—I have read and understood the permit and accept the precautions to be taken.

Signed   Position

1. _____   _____
2. _____   _____
3. _____   _____
4. _____   _____
5. _____   _____
6. _____   _____

10. WORK COMPLETED/PERMIT WITHDRAWN
Signed _____ Time _____ Date _____

11. ACCEPTED AND PERMIT CANCELED
Signed _____ Time _____ Date _____

12. COUNTERSIGNATURES			AWARENESS OF WORK IN ADJACENT AREAS
Endorsement	Plant manager or nominee	Appointed member of safety section	
Entry into confined space			1. _____ 2. _____
Source of ignition	_____	_____	Authorized Authorized
Renewal permitted	_____	_____	signatory signatory

RENEWALS
The work has been surveyed and conditions have not changed. Validity extended until:

Time	Date	Signature

GAS TEST RENEWAL   NOTE Exact positions to be clearly defined to eliminate any doubt.

The locations to be gas tested are	for

The following persons are authorized to witness repeat gas test associated with this permit:	
1.	
2.	
3.	
4.	

FIGURE 8-10. Sample work permit form.

Properly prepared permit forms promote safe working practices, but they cannot replace careful consideration of the risk involved for every occasion. Using check lists and similar tools as part of the printed form is a matter of individual company choice. When preparing a work permit, the effects of the work on adjoining parts of the plant, or even off-site effects, should be considered. The content of a work permit form should remind the signatories of their responsibilities and of the importance of the document to which they are a party. The requester should fully understand the nature of the work to be done and the precautions to be taken when he accepts the form. Figure 8-10 is a sample work permit form.

The work permit system should provide for stopping work for reasons other than its completion and for normal activities to resume if the work cannot be completed in the time allowed. A single work permit issuing point should be designated. The number of forms available to each department should be limited and they should be used in sequential order. A record showing the department to which the work orders were issued and the dates of their issue and return should be kept. A work permit system needs both unscheduled checks and continuous monitoring. If this is done properly and with the support of management, the scheme should operate effectively. It is also worthwhile for someone outside the department to audit the system from time to time.

Too much elaboration should be avoided when developing a work permit system. The system should be clearly understood, unambiguous, and written in a practical way. Unnecessarily bureaucratic procedures should be avoided. Since first- and second-line maintenance supervisors are responsible for making the work permit system work, they are in an excellent position to suggest modifications that will make the system work more effectively.

8.9. MAINTENANCE MANAGEMENT INFORMATION SYSTEMS

Incorrect maintenance information can also create hazardous situations. The chemical process industry literature[13] contains many examples of catastrophic failure that can be traced to maintenance activities that relied on incorrect information. Cutting the wrong pipe, using the wrong replacement part, deenergizing the wrong electrical circuit, using the wrong materials and wrong procedures are examples of applying incorrect maintenance information. Use of the wrong materials can lead to process safety incidents with dire consequences, as shown in Figure 8-11.

8.9.1. Work Order Tracking

For the daily monitoring of work in progress, a project tracking system should be in place. The maintenance department's need to respond to emergency

---

Factual Description:

In a crude production facility, a weld on a 12" pipe containing heavy hydrocarbons at high pressure and high temperature was identified as close to failure. The section of the pipe was isolated and the pipe was rewelded; however, the new weld failed the post weld inspection. A decision was made to replace a section of the pipe. Because proper materials control and an identification system were lacking, a section of pipe was fabricated from incorrect materials and installed. After the welding was completed and the weld tests were passed, the insulation was put back in place and the pipe was brought back into service. Two days later the new section of pipe failed releasing a large quantity of flammable hydrocarbons. The ensuing vapor cloud ignited, causing extensive property damage and seriously injuring several workers.

Contributing Causes	Process safety concept
1. Improper material control.	Equipment integrity, Maintenance procedures
2. Improper material identification.	Standards, Documentation, Maintenance procedures

FIGURE 8-11. Improper material control incident.

repairs requires a flexible schedule system. The usual method for beginning maintenance is to issue a work order. Before computers, multiple copies of work orders were needed for various craft, clerical, and supervisory personnel. Now the only printed copy of the work order may go to the maintenance mechanic, and the parts department may receive a parts and supplies list linked to the work order itself.

Work order tracking systems are particularly useful when multiple, linked work orders or sequential work orders are in place. Sometimes work orders must be suspended rather than closed, usually because parts or specialized tools are not available or because higher-priority work must be done first. The suspended work orders remain on the outstanding list and can be reissued when the work becomes feasible. Incorrect use of the work order system can lead to catastrophic accidents. The Piper Alpha disaster discussed in detail in Section 8.2.4 is a vivid example of the potential results of the work order system failing.

When a work order request has been completed, the maintenance data should be entered into a computer. Some software systems allow this information to be entered by clerks, who obtain the data from notations made by repair personnel on the work order. Other software systems are set up so that the maintenance technician enters the information directly.

8.9.2. Process Equipment Files

Process equipment files contain important information needed by the maintenance department. Several important categories of information are typically kept on each type of process unit. Besides the data supplied by the manufacturer and the process design team, a history of the equipment is needed for proper maintenance. Categories of information in the process equipment file are:

- Design/fabrication information
- Maintenance procedures
- Repair history
- Inspection procedures
- Inspection and test records
- Spare parts
- Special tools required.

Additional data can be recorded for certain types of process units, such as the historical trend of selected monitoring instrumentation or time spent on line. Some maintenance departments also accumulate cost-of-repair information.

Although a process equipment file generally focuses on a particular unit, because common components are used on many different types of equipment, certain information should be selected from each process equipment file and assembled. Examples are pressure relief valves and temperature sensors. Although pressure relief valves are associated with a particular pressure vessel or piping segment, the maintenance staff would find a cross-referenced relief valve list convenient. The need to be able to gather interrelated data contained in many different process equipment files complicates the maintenance department's task and can lead to accidents, as shown in Figure 8-12.

8.9.3. Process and Equipment Drawings

Before critical maintenance work begins it is essential to review the piping and instrument diagrams (P&IDs) to ensure that the section of the plant involved in maintenance activities is isolated from the operational portions of the plant. It is therefore critical that the P&IDs are kept up to date. Besides P&IDs, equipment drawings are needed to ensure that the correct part of the equipment is being subjected to the maintenance procedure. If a pressure vessel drawing shows only one inlet and one outlet, but the on-site maintenance crew

---

Factual Description:
An example of an accident caused by erroneous maintenance information involved the maintenance of pressure relief valves. Two pressure vessels, with similar identification numbers, had the same type of relief valves, but with different pressure relief settings. During the calibration of one relief valve, the higher pressure setting, erroneously read from the maintenance files for the wrong vessel, was used. The error was not discovered until a few years later when the vessel was damaged by unrelieved overpressure.

Contributing Causes	**Process safety concept**
1. Improper equipment identification, incomplete equipment files.	Standards and codes, Documentation

---

FIGURE 8-12. Erroneous maintenance information incident.

Factual description:
Several workers were seriously injured by acid ejected from a pump on which work was being done. The pump had been isolated with blinds, but they were not effective because of their placement. The work permit had properly required that the pump be isolated, but it did not specify where the blinds should be inserted, and no P&ID was available to review. Therefore, the lack of proper maintenance information was responsible for this incident.

Contributing causes	Process safety concept
1. Failure to keep P&IDs complete and up-to-date.	Process knowledge, Identification
2. Failure to give specific details about the placement of blinds in the work order.	Communication, Maintenance procedures

FIGURE 8-13. Erroneous blind placement incident.

discovers a new line to the vessel that is not on the equipment drawing, maintenance work should not proceed until the discrepancy is evaluated and resolved. Figure 8-13 describes an incident caused, in part, by not having complete, up-to-date P&IDs.

Piping and instrument drawings sometimes are not really detailed enough to show the entire protective system installed in the plant. A link from a sensor to a final control element may go through a series of devices before the simple control function is accomplished. The exact procedures for maintaining computer-controlled process equipment should be documented.

Another critical set of drawings that should be maintained in an "as built" state are process control loop diagrams. It is important that maintenance people work on the correct electrical circuit when maintenance work is being tested on process control system. Incorrect loop diagrams can cause extensive delays in the maintenance effort. Figure 8-14 describes an incident caused by improper labeling.

Factual description:
A vessel high-level alarm was replaced by maintenance personnel; however, the proof testing of that high-level alarm was not correctly done because the connections in the junction box were incorrectly labeled. The testing personnel therefore tested the working low-level circuitry. The new high- level alarm was defective, but it was not tested because of the loop diagram error. Several months later a process excursion, not recognized by operating personnel, eventually led to the vessel overflowing. During the subsequent incident investigation, the defective high-level alarm was discovered and the erroneous loop diagram was declared the root cause of the incident.

Contributing causes	Process safety concept
1. Incorrect labeling of connections.	Documentation, Maintenance procedures
2. Installation of defective equipment.	Installation procedures

.FIGURE 8-14. Improper proof testing incident.

Maintenance personnel should evaluate the need to update the P&IDs after any plant modifications they make. They should also determine the need for subjecting the change to management of change procedures.

8.10. QUALITY CONTROL

Preserving the quality of materials and workmanship during maintenance activities is very important to process safety objectives. A formal quality control program should be an integral part of any maintenance organization. Such a program typically covers procedures for ensuring that proper replacement parts are used and that plant equipment is inspected, as well as maintenance workmanship issues.

8.10.1. Replacement Parts

The proper selection of replacement parts is critical to safe process operation. Replacement parts should be stored in warehouses and controlled to ensure the selection of proper parts by maintenance personnel. The maintenance organization controls quality in this area by taking systematic action to confirm that all the requirements for replacement parts supplied by equipment vendors and contractors are met. These actions may include inspecting one or more parts manufacturing stages and auditing the effectiveness of vendors' and contractors' internal quality control procedures. The amount of verification required depends on the criticality, or the loss potential, of the replacement part in the chemical process.

The International Organization for Standardization (ISO) has published a series of standards, ISO 9000–9004 for quality systems. These standards describe, in general terms, the management systems and procedures that an organization should conduct to operate effectively and to give management and external bodies confidence in the organization's performance. The existence of an ISO 9000 series quality program at an equipment supplier or vendor does not completely guarantee the quality of replacement parts but it does mean that the quality control system at the supplier's location may be easier to audit and inspect. For this reason, many chemical plants have adopted programs that incorporate some of the ideas in the ISO 9000 standards.

The purchase of spare parts from sources other than the original equipment manufacturer (OEM) should be approved and documented. Using parts other than OEM parts requires a management of change review, which is described in Section 2.12 of Chapter 2, to ensure that the replacement part meets or exceeds the original design intent. The use of non-OEM parts may require pilot testing and close quality and performance monitoring during initial operation.

The use of OEM parts however, does not guarantee that the replacement part is exactly equivalent to the parts purchased formerly. Equipment manufacturers sometimes make changes that affect the integrity of the part in the chemical process service. A replacement material in a seal part, for example, may have different characteristics with respect to its reaction to chemical contact. Recently, a potentially serious incident was attributed to the replacement of a seal in a valve used to divert process fluid under conditions of impending overpressure. The valve would not open because the new seal material had swelled in contact with the process materials. The manufacturer had not informed the chemical company of the change in seal material nor had the user compared the new valve component list to the old list.

Some safety-critical equipment may require certified spare parts or replacement equipment certified by third parties. Dry chemical fire extinguishers are, for example, certified safety equipment and should be repaired by certified companies or technicians.

When used equipment is installed, the integrity of the equipment should be checked to determine if it is adequate for the intended service. Procedures for determining pressure vessel fitness for service are developed on a case-by-case basis, following careful inspection and analysis.

8.10.2. Inspection

A detailed equipment register and recording system are vital to maintenance and inspection for ensuring that items are not overlooked during inspections and that the lessons from maintenance history are heeded. Without scheduled inspections and prompt maintenance, chemical plant equipment and machinery can lapse into a dangerous state from wear, fatigue, or corrosion. Regular inspections are essential to determining the extent of deterioration. Since the inspection results may call for shutting down the plant, the staff responsible for making inspections should, if possible, be insulated from the operational organization. The person responsible for planning inspections in any organization may be given the authority to overrule production pressures, especially when the safety of personnel and the integrity of the plant are clearly at stake. In many small- and medium-sized organizations, maintenance and inspection functions are combined in a single department. In larger organizations, a separate, independent inspection department is common. For hazardous processes, a separate inspection department may be needed.

Inspection starts with the materials and equipment ordered for construction and continues throughout the life of the plant. Some inspections have to be done by maintenance workers, even when there is a separate inspection department. Most maintenance and inspections require that the plant and equipment be shut down and prepared. Condition monitoring is an inspection technique used while the plant is running. A few inspection jobs can also be done while the plant is running. Various mandatory inspections can be carried out by specialist companies that follow regulatory requirements for cranes, boilers, and other specialty equipment. Insurance carriers often require special third-party inspections of equipment such as fire water pumps and some safety equipment.

8.10.3. Certified Equipment

Because of the potential for safety-protective devices to fail and because of pressure from the public, there has been a growing trend in recent years for regulatory agencies to become involved in analyzing the reliability of protective systems. This has led to various approaches to certifying process equipment[12]. Historically, the National Fire Protection Association (NFPA) has published rules covering fire extinguishers, boiler control interlocks, etc. The insurance industry has promoted many voluntary codes that require equipment to be certified by third-party testing groups such as Factory Mutual (FM) or Underwriters Laboratory (UL). In Germany, all safety systems must be certified by Technischer Uberwachsungs Verein (TUV) before they can be used in applications critical to safety. The TUV certificate is recognized throughout the world, and many companies specify that only TUV-certified safety systems can be used in their safety-critical processes. Several other countries have enacted regulations to ensure that high-integrity safety systems are used. Technical societies, including the International Electrotechnical Commission (IEC), the Instrument Society of America (ISA), and the German Electrotechnical Commission (DINVDE), are all implementing new and increasingly stringent standards for safety-protective systems. In the United Kingdom, The Health and Safety Executive (HSE), is monitoring high-integrity programmable electronic safety systems.

8.10.4. Continuous Improvement

Continuous improvement is a quality improvement strategy that many companies are adopting in today's competitive marketplace. Continuous improvement is based on the belief that to be the best quality provider of goods and services, an organization must continually improve. Maintenance groups are included in this plan for continual improvement.

Continuous improvement focuses on a cycle, sometimes called the "Plan–Do–Check–Act" cycle. The first step of the cycle identifies a specific improvement opportunity and then develops a plan for testing the improvement. Analysis and solution development techniques include objective definition, worksheets, brainstorming, scatter plots, control/run charts, flow charts, cause and effect analysis/fishbone diagrams, fault tree analysis, multivoting, solutions matrix, and Pareto charts.

8.11. CONTRACTOR SAFETY

When hiring contractors to do maintenance work, the employer should ensure that the contractor has the appropriate job skills, knowledge, and certification (such as pressure vessel welders need). Since the contractors may be working in and around processes that involve hazardous chemicals, the contractors should also be selected for their past experience performing the desired tasks without compromising the safety of employees at the plant. Workers should be made aware of the hazards at the plant where they are working. They should be familiar with all relevant safety procedures, and special consideration should be given to potentially hazardous operations.

To help managers keep in touch with the whereabouts of all the maintenance workers and contractors in a process area, a "see at a glance" method display system can be used to identify the various types of activity and operations. The display should be located in the plant control room or in a similar central location. Maintaining a site entry and exit log for contractors is another way employers can track and keep current knowledge of activities involving contract employees working on or near a process. Injury and illness logs of both the employer's and the contractor's employees allow an employer to have full knowledge of process injury and illness experience.

First- and second-line maintenance supervisors may be responsible for managing any contractor performing maintenance and, therefore, should be aware of company policy and government regulations covering contractor safety. Government regulations in the United States identify the responsibilities employers have for contractors involved in maintenance, repair, turnaround, major renovation or specialty work on or near covered processes. The regulations require employers to: (a) consider safety records when selecting contractors, (b) inform contractors of potential process hazards, (c) explain the facility's emergency response plan, (d) develop safe-work practices for contractors in process areas, (e) evaluate contractor safety performance, and (f) maintain an injury/illness log for contractors working in process areas.[1] The regulations also require contractor employers to train their employees in safe-work practices and document that training, ensure that employees know about any potential process hazards and the employer's emergency action plan, ensure that employees follow the safety rules of the facility, and advise the employer of any hazards the contract work poses or hazards identified by the contract employees.

In addition to complying with the regulations, the main concern of maintenance personnel is to ensure that the contractors working in the plant do not endanger themselves or plant personnel. Close monitoring of contractor mainte-

---

1 Specific guidelines for using contractors are given in U.S. Department of Labor regulation [29 CFR 1910.119 (h)]. A summary of the requirements appears in Appendix A.

2

8. MAINTENANCE

TABLE 8-3
Contractor Safety Items

1. Check safety records before hiring contractor.
2. Make contractor aware of hazards at the plant.
3. Review safe-work practices with contractor.
4. Give special consideration to potentially hazardous operations when contractors are present.
5. Explain emergency response plan to contractor.
6. Provide a map showing the whereabouts of maintenance personnel and contractors by location and activity.
7. Maintain a site entry and exit log.
8. Maintain an injury and illness log.
9. Monitor and evaluate contractor safety performance.

---

nance activities will lessen the chance that an incident will occur. A summary of actions that maintenance supervisors can take to ensure contractor safety appears in Table 8-3.

8.12. INCIDENT INVESTIGATION

In the context of process safety, "incidents" include fires, explosions, releases of hazardous substances, or any event that results in death or injury, adverse human health effects, or environmental or property damage. Near misses are events that could have resulted in an accident or incident. The investigation of incidents and near misses is required by regulation and should be required by policy. First- and second-line maintenance supervisors are responsible for reporting any incident or near miss that occurs during their watch or that involves any maintenance work they are overseeing.

The maintenance department is a vital resource in any process incident investigation. The history of the process units, including repair records, previous outage incidents, and maintenance procedures, is included in the fact gathering portion of most investigations. If the incident occurred during a maintenance activity, maintenance staff may be included on the investigation team to provide important insight into the status of the process units. Communication among the members of an investigation team is critical to the success of their effort. The role of maintenance staff in investigating incidents, either as team members or as providers of information, should be to give their perspective of the events or of the circumstances that led up to the incident.

All maintenance work requires cooperation and communication with operations staff. Complete incident investigations usually reveal multiple causes. To determine the causes and make needed changes to prevent a recurrence of the incident or near miss, all resources should be used. It is not possible to separate the roles of the maintenance department from those of the operations department during an incident investigation. A detailed discussion of this is contained in Section 7.8 in this book. First- and second-line maintenance supervisors should read Section 7.8 and be familiar with its contents.

A recent major accident highlights the role of maintenance staff during in the incident investigation. The July 5, 1990, explosion of a waste water tank at the ARCO chemical plant in Channelview, Texas[15] occurred during the repair of a vapor compressor (also see Section 7.2.1). Information from several maintenance groups was needed to discover the various root causes of this tragic event, which killed 17 workers. The vapor compressor had failed multiple times in a short period. Information from the plant mechanical repair staff, the plant electrical staff (who were involved in a related control panel modification), and the vendor who reconditioned the unit was needed to establish the status of the tank and the vapor compressor subsystems. A tank oxygen sensor had been found inoperative, meaning that information was needed from the instrument maintenance group. An on-site contract maintenance company, which had craftsmen assigned to the vapor compressor repair project, also gave relevant information to the investigation team.

Since the main objective of any incident investigation is to learn from past mistakes, the lessons learned should be shared with all maintenance personnel. Awareness of the importance of maintenance activities to the process safety achieved at any chemical process facility is reinforced by periodically reviewing incident case histories (also see Section 7.8.3 for a discussion on effective use of incident investigation in training).

8.13. SUMMARY

The main objective of the maintenance department is to make repairs and to do preventive and predictive maintenance to maintain the mechanical integrity of the plant. Any equipment used to handle or process hazardous materials should be designed, built, installed, operated, and maintained in a way to control the risk of releases or accidents. The first responsibility in the maintenance of a plant is to maintain the design conditions of protective systems and not compromise them either temporarily or permanently. When installing new or when modifying equipment, the designed-in protection should not be compromised. When performing repairs or preventive maintenance, first- and second-line maintenance supervisors should ensure that proper preparations are made to protect their staff, the plant, and others while the work is performed and that the equipment is returned to service with all safety-protective elements in place. During maintenance, maintenance procedures, the work permit system, and safe work practices should be strictly followed.

First- and second-line maintenance supervisors should enforce and monitor all maintenance safety procedures. It is their responsibility to make sure that their staff have the appropriate training to understand the maintenance procedures, the safe practices, and the proper use of equipment and tools. They should also be alert to any changes that require management of change procedures. All maintenance work should be performed in cooperation and communication with operations staff. Once a job is completed, steps should be taken to communicate to those who need to know that the job is completed and that the equipment can be safely returned to service.

8.14. REFERENCES

1. Thornton, J. F. "PM—Key to Equipment Control." *Maintenance Supervisor's Handbook,* Edited by Frank L. Evans. Gulf Publishing Company, Houston, Texas, 1962.
2. Evans, F. L. "Preventive Maintenance in Use." *Maintenance Supervisor's Handbook,* Edited by Frank L. Evans. Gulf Publishing Company, Houston, Texas, 1962.
3. Cullen, W. D. *The Public Inquiry into the Piper Alpha Disaster.* Her Majesty's Stationary Office, London, November 1990.
4. National Association of Corrosion Engineers. *Publication 6H189: A State of the Art Report of Protective Coatings for Carbon Steel and Austenitic Stainless Steel Surfaces under Thermal Installations and Cementations Fireproofing,* Corrosion Engineering Task Group 7-6h-31, 1989.
5. American Petroleum Institute. *Recommended Practices 570: Inspection, Repair, Alteration and Rerating of In-Service Piping Systems,* 1st edition. Washington, D.C., 1992.
6. American Petroleum Institute. *Recommended Practice 653: Tank Inspection, Repair, Alteration, and Reconstruction.* Washington D.C., First edition, 1991.
7. Barkley, D. A. "Protecting Process Safety Interlocks." *Chemical Engineering Progress,* February, 1988.
8. Knee, H. E. "Human Factors Engineering; A Key Element of Instrumentation and Control System Design." Proceedings of the 39th International Instrumentation Symposium. Pater No. 93-146, Albuquerque, New Mexico, May 2-6, 1993.
9. Kletz, T. A. "Three Mile Island: Lessons for HPI." *Hydrocarbon Processing,* June, 1982, pp. 187–192.
10. McMillan, G. K. "Can You Say Process Interlock?" *Technical Journal,* May 1987.
11. Kletz, T. A. "A Three-Pronged Approach to Plant Modifications." *Loss Prevention, Volume 10, Chemical Engineering Progress.* American Institute of Chemical Engineers, 1976.
12. Townsend, A. *Maintenance of Process Plant—A Guide to Safe Practices,* 2nd Edition. Institution of Chemical Engineers, London, 1992.
13. M&M Protection Consultants. *One-hundred Largest Losses: A Thirty Year Review of Property Damage Losses in the Hydrocarbon-Chemical Industries,* Ninth Edition.
14. Zodeh, O. M., Sikora, D.S. "Self Checking Safety Interlock Systems." *IEEE Transaction Industry Applications, Volume 25, #5,* September 1989
15. *Guidelines for Investigating Chemical Process Accidents.* American Institute of Chemical Engineers, 1992.

Additional References

Systems Reliability Directorate. *Pipework Failures, a Review of Historical Data.* SRD-R-441, pp. 12, UKAEA, Wigshaw Lane, Culcheth, Warrington, WA3-4NE, United Kingdom, 1987.

King, Ralph. *Safety in the Process Industries.* Butterworth-Heinemann, London, 1990.

Richmond, D. "Instrumentation for Safe Operation." *Safety and Accident Prevention in Chemical Operation,* edited by Occit and Wood. Wiley-Interscience, New York, 1982.

Health & Safety Executive. *PES-Programmable Electronic Systems in Safety Related Applications.* Her Majesty's Stationery Office, London, 1987.

Wells, G. L. *Safety and Process Design.* Godwin, London, 1980.

Marshall, V. C. " What Happened at Harrisburg?" *Chemical Engineer,* 346, p. 479, London, 1979.

Sanders, Roy E. *Management of Change in Chemical Plants: Learning from Case Histories.* Butterworth-Heinemann, London, 1993.

9

SHUTDOWN

Since shutdown is not a routine operation, the potential for process safety problems is greater during this time than during normal operations. A recent analysis[1] of catastrophic accidents in the hydrocarbon and chemical industry showed that an accident is five times more likely to occur during "other than normal" operational status, including during shutdown and startup.

Shutdowns are necessary for many reasons, ranging from a scheduled event to an emergency condition. The operations staff defines shutdown as the sequence of steps used to bring a chemical process from operational to nonoperational status. The maintenance staff defines shutdown as the nonoperational period before startup, when they perform repairs and maintenance functions. Shutdown also is defined by specific time periods and the events that occur within those periods. The shutdown time periods are shown graphically in Figure 9-1 and are listed here:

- Pre-shutdown planning (during operating period)
- Shutdown sequence
  —initiation
  —execution
  —conclusion
- Shutdown (a stable nonoperational state)

For a continuous process, the shutdown may be a scheduled or an unscheduled event that occurs infrequently in response to either maintenance or emergency situations. For a batch process, the shutdown sequence is the method for stopping the normal planned batch procedure.

The initiation and conclusion of the shutdown sequence are specific milestone events. The time needed to conduct the shutdown sequence varies from almost instantaneous to several days, depending on the specific chemical process involved. The pre-shutdown and nonoperational shutdown periods also vary, depending on the process and the goals of the shutdown. Accidents can occur because of the unfamiliarity of operations personnel with the shutdown sequence. During the nonoperational period, accidents may occur

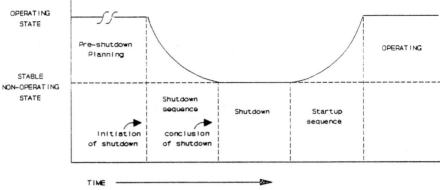

FIGURE 9-1. Shutdown periods.

because the manufacturing staff is under pressure to return to operational status. The purpose of a shutdown is to place the process in a stable nonoperating state. All chemical processes, both batch and continuous operations, must eventually experience a shutdown. Three categories of shutdown—normal, extended, and emergency—are defined below:

- *Normal Shutdown:* The chemical production facility is taken out of service in a scheduled routine way for maintenance, process modifications, inventory control, end of batch, etc.
- *Extended or Mothball Shutdown:* Extended shutdown is a special case of normal shutdown, caused by equipment damage, lack of product demand, or other business reasons; the intent is to restart the unit at some future date.
- *Sudden or Emergency Shutdown:* An unplanned shutdown can be caused by a critical process parameter exceeding its preset limit, an unexpected equipment failure, or utilities failure.

The distinguishing factor in the above categories is the *scheduled* nature of the shutdown. Normal shutdowns and extended or mothball shutdowns are planned events with a schedule that can be distributed to all participants. The sudden or emergency shutdown is an unscheduled event that depends on the preplanning and training of operations staff and specialists in emergency response.

The operations staff is responsible for performing the shutdown sequence; however, maintenance staff participation is required during all the shutdown phases. For example, maintenance staff may be called on to provide temporary piping for inventory reduction and decontamination. A clear definition of the responsibilities of each group and methods for communicating should be established in the written shutdown procedures, which should also define the "ownership" responsibility for the equipment during the various phases of the shutdown. Responsibility for equipment or facilities may shift from operations to maintenance or to other groups, such as the fire brigade, specialist contractors, or outside authorities. The exact time or point when responsibility shifts should be specified in the written shutdown procedures.

Although it is desirable to have standard shutdown procedures throughout all facilities, some processes or locations may need special techniques. A requirement to reduce operating inventory (dump production to the flare, alternate containment vessels, or the sewer; or other appropriate techniques) is common to most shutdown procedures. For some chemical processes, partial or localized inventory reduction procedures may be appropriate. Other common shutdown procedures, such as isolation, purging, and inerting, are specific to each process area.

Conducting a process hazards analysis of all shutdown procedures is an important part of a process safety management program. It is especially important to conduct a process hazards analysis of any changes to the general shutdown procedures particular to a specific process area, or of changes due to special circumstances, such as the unavailability of equipment or an unusual event.

9.1. NORMAL SHUTDOWN

A normal shutdown is a scheduled event. The schedule for a normal shutdown is determined by the required inspections, preventive maintenance, equipment repairs, and process modifications. A specialized instance of a normal shutdown is the mothballing of a facility. A shutdown can range from a very simple operation, such as taking one of several parallel pumps out of service for routine maintenance or minor repair, to a complex multiweek plant turnaround. For a turnaround, the shutdown sequence will include removing inventory and isolating and decontaminating vessels and equipment to prepare for maintenance. It may be necessary for the maintenance staff to help the operations staff during the shutdown sequence. Such assistance could include providing temporary piping for draining and decontamination, purging, and installing blinds and valves. The temporary placement of a specific process unit in an off-line mode for quick repair is within the context of a normal shutdown. Placing specific instrument sensors in an off-line or manual mode for repair or calibration while the process continues is *not* considered a shutdown.

Each chemical process has specific procedures for a shutdown. The topics in the following paragraphs address general process and personnel safety factors that should be considered for a normal shutdown.

9.1.1 . Pre-shutdown Planning

Shutdown planning has two major overlapping and tightly interrelated elements: planning the steps of the shutdown sequence, and planning the work activities during each shutdown period of the shutdown sequence.

Operations personnel or the associated production technical staff generally are responsible for planning the shutdown sequence. When safety protective systems should be tested during the shutdown sequence, close collaboration with the maintenance group is necessary. Also, it is the operation department's responsibility to ensure that the process remains shut down.

Planning the work flow during the shutdown period is the responsibility of the maintenance group. For a typical maintenance shutdown, the operations staff generates a "job list" from work orders. The work orders are submitted by both operations and maintenance staff who specify the equipment in need of repair. The maintenance planners should include these items in their scheduled maintenance tasks.

Communication between the maintenance and operations planners is critical at this stage, because the safe performance of the shutdown is a result of their cooperative efforts. Miscommunication at this time can cause delays or even a hazardous incident. Many reported incidents[2] can be attributed to communication problems, which can lead to inadequate on-site materials to perform the maintenance, fuzzy task responsibility assignments, or inadequate time to perform a maintenance task.

9.1.1.1. Process Safety Considerations

A written procedure should be available for each shutdown sequence step or maintenance procedure during the shutdown period. A process hazards analysis document should be referred to or incorporated into the written procedure. It is the responsibility of the shutdown planners to verify the relevance of the process hazards analysis. If conditions or equipment have changed, rendering the process hazards analysis incomplete or inadequate, the shutdown planners should obtain an updated process hazards analysis before completing the shutdown plan. An effective way to ensure the adequate consideration of process safety issues during the shutdown preplanning activities is to use management of change procedures where appropriate.

In some chemical processes, an important safety consideration of the shutdown sequence is to avoid the flammable range during the removal of inventory or the decontamination of equipment. If possible, shutdown procedures should not take the process through the flammable range. An example of this situation is during the manufacture of acrylic acid, when it becomes necessary to shut down the reactor. The reactor normally operates with an air–propylene mixture above the upper flammable limit. Steam is added to the reactor to maintain the reactor contents outside of the flammable region throughout the shutdown, as shown in Figure 9-2. The shutdown sequence

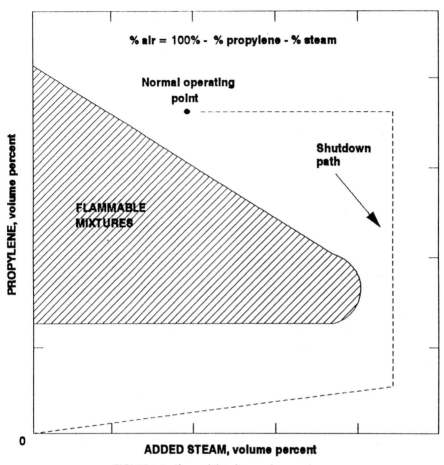

FIGURE 9-2. Flammability diagram for propylene.

path in Figure 9-2 shows how the flammable region is avoided by replacing air with steam until the mixture in the reactor is beyond the flammable range, then reducing the propylene feed percentage until the mixture in the reactor is below the flammable region. Finally the steam can be shut off and the reactor purged with air. A simpler and more "closer to home" example is the case of shutdown and emptying of a tank containing a flammable liquid—be sure to avoid the flammable condition in the tank by using inert gas rather than air as the tank is emptied—and introduce air only after the tank is emptied. This is treated in more detail in Chapter 10.

9.1.1.2. Job Task Analysis
To ensure the safe execution of both the shutdown sequence and maintenance or repair during the shutdown period, each task within each job should be

analyzed for process safety considerations during the pre-shutdown planning period. Jobs are defined as a set of tasks needed to accomplish the job goals. The major features of a job task analysis are:

- Steps required to accomplish the work
- Tools and equipment resources
- Man–machine interfaces
- Knowledge, skills, and abilities of the work crew
- Milestones, cues, and targets
- Expected process environmental conditions
- Applicable process hazards analysis
- The difficulty rating of each task
- Document control, change control procedures
- Task validation requirements

When these features have been addressed, the shutdown planners can forecast their resource needs. Before they develop a schedule, these resource needs are matched with available resources (personnel, equipment, tools, process or equipment timing constraints, etc.).

If the shutdown includes specialty repairs, unique maintenance requiring outside contractors, or process modifications by a construction contractor, further job task analyses are needed before the shutdown plan is completed. When maintenance or repairs require the use of outside contractors, the job lists should be augmented with additional items according to company guidelines and OSHA requirements. These additional considerations should include:

- Examine contractor's safety record
- Inform contractor of potential process hazards
- Inform contractor of facility's emergency response plan
- Develop safe work practices for contractors in process area
- Evaluate contractor's safety performance periodically throughout the job
- Maintain an injury/illness log for contractors working in process areas

The operations and maintenance departments play an integral role in analyzing job tasks since they are the stewards of the local site-specific conditions that greatly affect the analysis.

9.1.1.3. Chemical Inventory Reduction

When planning a shutdown, the initial step of the shutdown procedure may be to reduce the inventory of process chemicals to the lowest safe level. Reducing inventory decreases the amount of material that must be managed during decontamination. Some chemical processes are mechanically configured to permit a shutdown without a substantial reduction of chemical inventory. Specific unit isolation capabilities may enable chemical inventory reductions to be avoided.

9.1.1.4. Isolation of Equipment for Maintenance

Preplanning may include procedures for isolating equipment scheduled for maintenance during the shutdown period. The planners should follow both company and regulatory guidelines regarding the safe isolation of equipment. The 1990 OSHA lockout/tagout standard (*29 CFR 1910.147*) refers to the requirement that workplace procedures must be used to prevent the unexpected energization of equipment that may cause damage or injury during maintenance activities. Lockout refers to the process of installing locks on circuit breakers and disconnecting switches, blind flanges, or block valves so that the equipment cannot be operated until the lock is removed. A lock authorization procedure is implied. Tagout refers to the signs or warning labels attached to an isolating device to warn others not to operate the tagged equipment. The idea of lockout/tagout originated with electrical devices, but has been expanded to cover any situation where operating a piece of equipment could cause injury, damage, or the release of a hazardous material.

Initiating equipment isolation is typically the responsibility of operations. Still, each maintenance crew should verify the isolation and attach its own tag to any device that may affect their work. Only when the maintenance crew finishes the task can their tag be removed from an isolating device. In many facilities, only the person who installed the tag can remove it. Operations staff should not attempt to restart equipment until all tags are properly removed. In one case, an accident occurred when the lockout/tagout procedures were not followed during a pump repair. An electrician had reinstalled a pump motor after repairs. He reconnected the electrical leads, removed his tag from the local start/stop switch, went to the motor control center and removed his tag and locks. Then he remembered that he had not checked the pump rotation. The electrician energized the system, went to the pump and "bumped" the motor to check rotation. He found the rotation to be wrong, which required him to switch the power leads around. He disconnected the leads inside the motor junction box without tagging the local stop/start switch or deenergizing the breakers in the motor control center. Finding that he did not have enough tape to finish wrapping the leads, he left the site to get more tape. The operator in the area came to check job progress, saw no tags, saw the breaker in the motor control center energized, *assumed* the work was complete and attempted to start the pump. The bare, unconnected leads arced, catching some nearby oil on fire.

Issuing confined space entry permits and hot work permits are operations tasks that depend heavily on equipment isolation, decontamination, and work procedures. Close cooperation among operations, maintenance, and specialty contractor personnel is needed to perform the specific maintenance and repair process vessels, piping, and equipment. The importance of the permitting procedure is shown by the following example. A tank cleaning contractor entered a tank without proper permits. The tank had been purged with nitrogen instead of air. Two workers entered the tank and were asphyxiated.

Operations should insist on the strict observance of proper permit procedures to avoid such tragic consequences.

9.1.1.5. Decontamination of Equipment for Maintenance

Decontamination is important in planning the final steps of the shutdown sequence, before the process equipment, piping, or vessels are transferred to the maintenance staff. Decontamination includes removing toxic and/or flammable materials. Careful planning is needed to find the number of vessel volume exchanges (length of time for a specific steam, water, or purge gas flow rate) needed to reach an acceptable level of decontamination. Both flammability and toxicity should be considered. Decontamination is typically performed by the operations staff, with assistance from the maintenance department. The operations department performs the final testing of equipment and issues the appropriate permits before turning the equipment over to the maintenance department. The operations department is responsible for ensuring that the equipment stays decontaminated for the maintenance period.

9.1.1.6. Regulatory Issues

For many shutdown situations, many regulatory issues should be addressed during the pre-shutdown planning period. The process safety, worker safety and environmental regulatory framework is intertwined in so many areas that a review of the shutdown (particularly a large unit shutdown) by safety and environmental specialists may be warranted.

The worker safety regulatory issues that affect the process safety considerations of a proposed shutdown are often explained in a Safe Work Practices Manual. This manual would normally contain specific forms and procedures used in the plant to accomplish the various work practices. Table 9-1 presents an example list of topics contained in a Safe Work Practices document.

The shutdown planners generally refer to the appropriate safe work practice guideline in the shutdown plan. Many of the safe work practice guidelines in Table 9-1 contain specialty checklists that help the planner allocate the proper resources to a task, as well as help the work crew perform the task. Table 9-2 is an example of a guidance list for using personal protection equipment.

A key environmental and regulatory issue affecting shutdown is the handling of gaseous, liquid, or solid waste that may be generated during the shutdown and decontamination sequence or because of the maintenance and repair activities during the shutdown period. Purge gases, particularly those containing flammable or toxic residues should be safely discarded through the flare or vent systems. Provisions for handling liquid or solid wastes can be a very complex environmental issue. The distinctions between hazardous and nonhazardous wastes, listed and characteristic wastes, threshold quantities, and the multitude of waste handling options require an analysis by an environmental specialist. The identification of waste handling tasks is usually a critical issue in most shutdown plans.

TABLE 9-1
Example List of Safe Work Practice Guidelines

Confined Space Entry	Hazard Communication
Machinery Guards	Vehicle Safety
Asbestos Maintenance	Lockout/Tagout
Contractor Safety	Hot Tapping
Flammable and Toxic Gas Detection	Record keeping
Cranes and Hoists	Ladders and Scaffolds
Electrical Safety	Personnel Protection Equipment
Excavation and Trenching	Respiratory Protection
Hot Work permits	Compressed Gas Cylinders
Static Electricity Prevention	Tank Cleaning
Fire Protection	Line/Vessel Entry ("First Break") Procedures/Permits

---

TABLE 9-2
Example Personal Protective Equipment Guidelines

The selection and proper use of personal protection equipment is critical to worker safety. Equipment to protect workers may be divided into four categories, according to the level of protection offered.

Level A   protection has a fully encapsulating chemical-resistant suit with self-contained breathing apparatus, double gloves, boots, hard hat, and a communication (two-way radio) system. Level A equipment is required when there is a high potential for skin, eye, respiratory, or gastrointestinal contact with hazardous materials.

Level B   protective equipment provides a high degree of respiratory protection, and a lesser degree of protection to the skin. Level B equipment should be selected when the risk of skin contact is not as severe as in Level A situations, but the risk of respiratory tract damage is high. The equipment has a self-contained breathing apparatus, chemical-resistant clothing, gloves, boots, and hard hat.

Level C   protection provides limited respiratory and skin protection. This level is selected when the conditions permit the wearing of air purifying devices. Conditions that permit the wearing of air purifying devices are:
- There is enough oxygen in the air (at least 19.5 percent).
- The hazardous substances have been identified and measured.
- The hazardous substances all have good warning properties.
- The device fits the face of the person wearing it.
- The appropriate cartridges or canisters are used, and the filtering capacity of these filters is not exceeded.

Level D   protection has coveralls, safety boots, safety glasses, gloves, and a hard hat. Level D is acceptable only when no hazardous air pollutants have been measured at the process unit site.

In some chemical processes, the available waste handling equipment may dictate the timing and sequence of the shutdown inventory depletion or decontamination procedure. Exceeding the waste handling capacity may not only be an environmental regulatory infraction but may lead to a serious process safety problem.

9.1.1.7. Management of Change Procedures

During the preplanning activities, situations may arise in which temporary piping or rental equipment must be used to accomplish the shutdown tasks. A management of change authorization procedure should be initiated to ensure that the safety implications of these unusual situations are carefully considered.

The Flixborough accident case is a classic example of the failure to determine the safety implications of temporary piping designed for shutdown purposes. In this case, an inadequate support structure (i.e., misapplication of flex bellows in something other than a straight run) was built for temporary process piping. When the structure collapsed, the chemicals escaped from the ruptured piping and the vapor cloud explosion caused widespread damage.[3]

Although most contingencies are considered when developing the shutdown plan, sometimes unexpected situations develop that require a deviation from the shutdown plan. Any deviation from the shutdown plan may require a management of change authorization procedure.

In one case, during the shutdown of a large propylene refrigeration unit, the shutdown and decontamination procedures required the installation of a 36" blind. The purpose of the blind was to isolate the two compressors not scheduled for maintenance so that a positive pressure with the process fluid could be maintained. When it was time to insert the blind flange (during the graveyard shift), no blind was available. The operations and maintenance personnel working the job decided to make a blind flange out of ¾" plywood. On repressuring the equipment to the planned pressure, the plywood blew out, allowing a large vapor cloud to form that was ignited by a passing vehicle. Proper management of change procedures would have revealed (through an engineering analysis of the capability of the plywood blind to contain the pressure) that a much lower positive pressure was needed when using the plywood flange or that the work would have to be delayed until a suitable part was obtained.

9.1.2. Shutdown Sequence Steps

Operations staff will always initiate the shutdown sequence, since it starts with an operating unit. For those processes operated continuously, shutdown is an infrequent event that may not be familiar to all operations staff, and unfamiliarity may introduce safety problems through human error. Another potential safety concern is the operation of equipment as it nears the shutdown state. Transients, both hydraulic and electric, may be experienced. Safety-protective systems may be inoperative, such as the low-level switch that is bypassed (jumpered) when a vessel is drained.

The operations department's responsibilities during the shutdown sequence typically include emptying, purging, cleaning, cooling, and verifying the safety of the nonoperational environment for the maintenance personnel. In a summary of accidents in the Hazard Loss Information System,[1] almost half the incidents were found to be due to deviations from the defined procedures. Clear, concise, written procedures for shutdowns are essential.

9.1.2.1. Written Procedures

Written procedures are particularly important for shutdown sequences and for maintenance or repair activities during the shutdown period. Written procedures, if developed using the job-task analysis techniques, should clearly identify the tasks and responsibilities of each person involved in the shutdown.

For example, a new instrument sensor must be placed in a pipe containing flammable material. Before welding a connection to the pipe, several preparatory steps should be taken. After the continuous flow has been rerouted via an existing bypass line, the pipe segment should be depressurized and flushed. Then a blind flange should be installed. Clearly the welder is not the appropriate person to be in charge of the preparatory steps. In some companies, operations staff install the blind flange, while maintenance staff have that responsibility in other organizations. A procedural flow chart, such as the example in Figure 9-3, identifies the steps of the procedure and the responsibilities of the various people involved in the preparatory task.

For extremely toxic or pyrophoric materials, a more elaborate preparatory procedure may be needed. A checklist such as that shown in Table 9-3 may be a part of the procedure. Throughout the shutdown, the operations department is accountable for the safety of the atmosphere in which the maintenance personnel must work.

9.1.2.2. Communications between Operations and Maintenance Departments

Proper communication between the operations and maintenance departments is important to executing the shutdown procedures safely. The operations department has ownership and control of the process area at all times, yet the operations department at various stages of the shutdown authorizes maintenance staff or others to perform work. The operations department is responsible for telling the maintenance department about the process area where work is to be performed. The maintenance department, on the other hand, should tell the operations department the nature of the work to be performed, the schedule, and the expected outcome. After work is completed, a summary

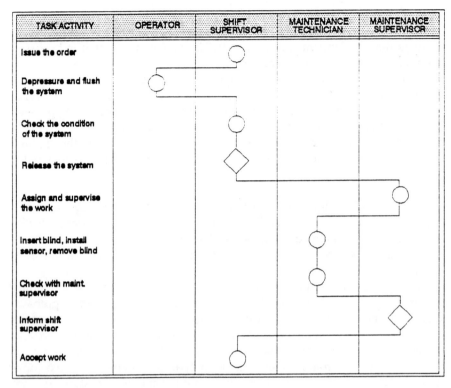

FIGURE 9-3. Example procedures used to replace a pressure sensor in process piping. O = Internal department action; ◇ = External department action.

TABLE 9-3
Example Checklist for Ambient Air Monitoring during Shutdown Maintenance

1. All lines blinded, isolated	9. Equipment grounding
2. Electrical locked out, tagged	10. Physical hazards
3. Vessels cleaned and purged	—Lighting
4. Proper ventilation	—Temperature
5. Atmosphere checked	—Scaffolding
—Lower Flammable Limit	11. Protective clothing
—Oxygen level	12. Protective equipment
—Toxic gas level	13. First aid kit/training
—Standby persons	14. Fire fighting equipment
6. Self-contained breathing	15. Communications
7. Apparatus at entry	16. Signs posted
8. Lifelines, rescue equipment	17. Preentry safety meeting and special precautions

of the accomplished work should be communicated to the operation department. In larger organizations, communications between the operations and maintenance departments may be by formal memoranda. In smaller facilities, formal exchanges of information may not be necessary since open channels of communication are more easily maintained.

Inadequate communication between operations and maintenance departments has led to accidents or emergencies. For example, in one location, dimethyl formamide (DMF) was being used to strip acetylene from an ethylene–acetylene mixture. As shown in Figure 9-4, the ethylene–acetylene mixture is fed to the absorber, where the DMF selectively absorbs the acetylene. The acetylene-rich stream goes to the stripper, where the acetylene is separated and the DMF is recycled to storage. To perform maintenance on the absorber and stripper, the unit had been shut down and temporary piping (shown by the dotted lines) was installed. Operations staff intended to use the temporary piping not only for decontamination but also to bypass the absorber for filling the stripper during startup. Unfortunately, the operations staff did not inform maintenance personnel of the dual purpose of the temporary piping. Maintenance staff performed the necessary work and removed the temporary piping after decontamination. Operations personnel opened the flow control valve to start filling the stripper, assuming the temporary piping was still there, and a large spill of the toxic material, DMF, resulted. This example shows the importance of clear communication between the operations and maintenance departments.

9.1.3. Testing Safety Protective Systems

Although most alarm circuits can be loop-checked during operation, some safety-protective devices can be tested only when the process unit is scheduled for shutdown. Some devices, such as gas detectors and pressure relief valves may be tested or calibrated off-line or during a maintenance shutdown period. Other devices, such as liquid level switches, can be tested during the shutdown sequence or during a specific portion of the shutdown period. For example, low liquid level switches in the lube oil reservoirs of expensive rotating equipment should be tested by lowering the liquid level, simulating actual service conditions. This test can be performed during the reservoir draining procedure when an oil change is needed. A liquid level switch should *not* be checked by simply lifting and lowering the float by hand.

For example, during the incident investigation following the failure of a large compressor due to a lack of adequate lube oil, a test of the lube oil reservoir low-liquid-level switch revealed that if the float were lifted by hand (as had been done during the most recent maintenance shutdown), it would

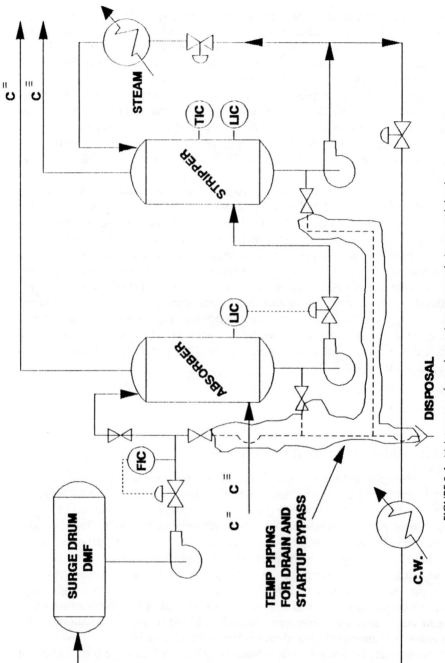

FIGURE 9-4. Maintenance of acetylene stripper system during normal shutdown.

activate the alarm. But when the liquid level decreased slowly over time, the switch would not activate the low-level alarm because of the mechanical looseness of the contact linkages.

Unlike alarm tests that may be performed during operations, the testing of most shutdown trip circuits should be scheduled for shutdown. Testing the safety protective trip circuits during a scheduled shutdown should be carefully considered instead of performing the conventional manual unit trip from the operator control panel. If a process unit has several trip circuits and is not scheduled for regular periodic shutdowns, priority should be given to trip circuits that are most vulnerable or most important.

9.1.4. Shutdown Period Maintenance Activities

The "ownership" responsibility for certain equipment (but not the area) changes from the operations department to the maintenance department during the shutdown period. Therefore, the maintenance department's responsibility for the process safety of the equipment (proper and safe repair) is the dominant factor during the shutdown period. The interfaces, such as the isolating flanges, tagged circuit breakers, piping spool pieces, etc., should be clearly understood by both operations and maintenance staff in order to define the area of maintenance responsibility. When only a portion or a single unit is involved in the shutdown (and adjacent equipment continues to operate), understanding between these groups about the ownership is vital. Chapter 8 contains examples of procedures, guidelines, and checklists used by maintenance for scheduled maintenance or repair during the shutdown period.

9.1.5. Unit Restart after Maintenance

After the maintenance department completes its various tasks on the shutdown equipment, operations staff can proceed with the restart procedures discussed in Section 6.5. The lockout/tagout practice would dictate that the operations staff not begin startup activities until all locks and tags are removed and inventoried.

9.1.6. Formal Review of Shutdown

A formal review of the shutdown should be conducted when possible (within a few days) after the shutdown ends. Such a review allows operations and maintenance personnel to critique each phase of the shutdown. Special emphasis should be devoted to process safety issues, and the review should focus on identifying problems that occurred and on developing solutions and improvements. The review results should be documented so that they can be incorporated in the planning process for the next shutdown.

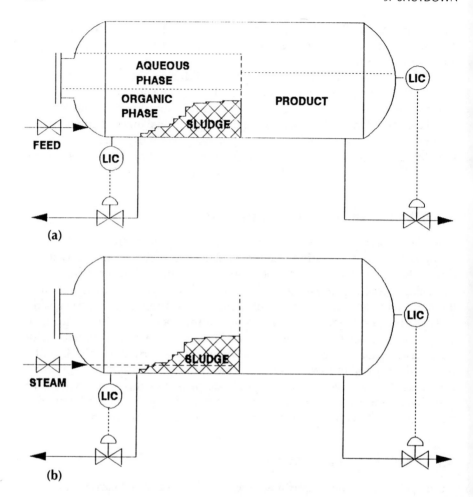

FIGURE 9-5. Decontamination of a baffled separator.

For example, a baffled separator was decontaminated by flowing hot water through the first section and out of the second section [see Figure 9-5(a)]. Repeated vapor tests showed that the vessel was still contaminated. The delay in decontaminating the vessel created scheduling problems, which could have had a later effect on process safety when the urgency to restart the unit resulted in unadvisable shortcuts. Sludge that had accumulated between the organic phase outlet and the internal baffle could not be removed by the water flush since too little volume or turbulence could be produced to remove it. During the review, it was decided that the next decontamination plan would call for draining the vessel and installing a lance through the inlet valve before beginning the water flush [see Figure 9-5(b)].

9.2. EXTENDED OR MOTHBALL SHUTDOWN

Extended or mothball shutdown is a special case of the normal shutdown. Additional safety concerns relate specifically to extended shutdowns, and these concerns increase the responsibility of both the operations and maintenance departments.[4] Simply stated, the goal of the mothballed shutdown involves removing process materials from the system and keeping them out.

Normal shutdown procedures may have to be changed to prepare the equipment for extended shutdown periods. For example, removing a piping spool piece may be a desirable isolation technique during the extended shutdown of a process unit. Although a double block and bleed piping isolation is acceptable for normal shutdowns in certain chemical process facilities, even a tiny, undetectable valve leak may cause safety problems in a mothballed unit.

In a recent accident, a worker was injured by directly contacting a phenol–water solution. The source of phenol in the mothballed unit was traced to a small leak in the double block and bleed that had accumulated in a downstream drip leg over many years until it spilled over into another connected vessel. During maintenance on the connected vessel, the worker was deluged with the phenol and seriously injured. Physically removing a piping spool piece would have prevented that particular accident.

Another precaution to take with a mothballed unit is to physically remove critical electrical equipment, such as entire circuit breaker boxes or termination panels. Simply removing the fuses or circuit breakers may not adequately protect mothballed units from future accidental reactivation.

For long shutdown periods, special corrosion protection, both internal and external, may be needed, and a corrosion monitoring program may be required. The corrosive states of plant equipment can be monitored periodically, typically using electrical potential techniques. Stress corrosion cracking, pitting corrosion, selective phase corrosion, and impingement attack can be monitored by these techniques. After purging, a positive nitrogen pressure may be used to reduce oxidation.

Freeze protection may be needed for instruments or special process units, to protect these units from undesirable or damaging temperature extremes. Written documentation of the mothballed shutdown and decontamination measures should be considered. This is important since different people may be responsible for restartup or dismantling.

Often a system of tags and other warnings may be used to label mothballed units. Monitoring barricades, warning tags, and other isolation devices are particularly important if the mothballed unit is next to or within the area of operating units.

9.3. SUDDEN OR EMERGENCY SHUTDOWN

Today's chemical process plants are highly complex integrated facilities with sophisticated compressors, pumps, turbines and process furnaces whose damage or destruction would severely affect the finances of a company or, in extreme cases, cause injury or death. Since some types of chemical process equipment can reach undesirable operating conditions rapidly, safety interlock systems are installed to shut down equipment before these unsafe limits are reached.

Operations staff are the key players in a sudden shutdown event, particularly in the initial stages of the shutdown sequence.[5] The maintenance department generally has little involvement, except to be members of the emergency response team or to get in a special operations role.

An example of a situation when maintenance personnel help operations staff during a shutdown is in a large manufacturing facility that has, for example, a large steam turbine used to drive a compressor, but the steam turbine is also part of the plant's steam distribution system. The inlet steam pressure to the turbine is 1500 psig and the turbine exhausts at 50 psig. The outlet lines from the turbine connect to the plant steam system at the battery limits. The lines are 24" and 36" and have manual block valves at the battery limit. During an emergency shutdown of the turbine, the battery limit valves have to be closed. Because not enough operators may be available to secure the unit, shift maintenance personnel can be assigned special operations functions such as closing the battery limit isolation valves for the turbine.

Safety interlock systems are critical to the safe and economic operation of process plants.[6] The safety interlock system can range from a simple shutdown valve or switch to a complex sophisticated computer system. Modern safety interlock systems can be constructed using several different types of technologies. There are the older "hardwired" relay-logic systems, and the newer solid-state logic systems that use programmable logic controllers. In these the "logic" is a computer program. Finally, there are systems that are a part of a complex Digital Control System.

All safety interlock systems share the requirement of a triggering event. The primary function of the safety interlock system is to execute the shutdown sequence for an operating process unit that has exceeded or threatens to exceed a predefined safe operating limit (speed, temperature, pressure, liquid level, vibration, etc.). Typical events that trigger or activate safety interlock systems include:

- Loss of power
- Loss of steam
- Loss of instrument air
- Loss of coolant
- Loss of fuel

- Loss of vacuum
- Loss of purge/ullage gas
- Operator error
- Critical instrument failure
- Safety protective device failure
- Rotating equipment failure
- Reactor runaway
- Loss of reaction

For simple chemical processes, one sudden shutdown sequence may be enough to bring the chemical process into a stable nonoperational state in response to any of the above safety interlock trigger events. For more complex processes, several sudden shutdown sequences may be required, some limited to single process units, others possibly affecting the entire facility.

Manual safety interlock procedures, in response to alarm signals or visual observations, are the simplest kind of sudden shutdown. Many older process facilities have pulldown charts in the control room to guide operators through the manual procedures needed to initiate a safety interlock action.[7]

Because of the increased use of computer technology, automatic shutdown sequences that respond to safety-protective instruments are now widespread. Combinations of automatic and manual shutdown sequences have also been installed that rely on both the installed safety-protective instrumentation and on manual intervention.

Figure 9-6 shows an example cause and effect diagram showing the detection and actuation (or response) of various emergency shutdown sequences in a complex plant. Each different safety interlock sequence is identified by a unique "trip" number. Trip number is a term that originated from the older manual shutdown systems that were activated by tripping a relay.

9.3.1. Preplanning for Sudden Shutdown

It is impossible to predict the timing and specific precursor conditions of an emergency shutdown. Therefore, to maintain control of the process and equipment during an emergency shutdown, preplanning and operations training is essential.[8–10] The planning usually has three distinct phases; (a) options, equipment, and safety-protective controls to avoid the shutdown, (b) if the shutdown is unavoidable, the sequence of necessary tasks and the specific responsibilities of the different personnel, and (c) after successful shutdown, the reviews and level of authorization needed to restart the unit.

Planning and training should be carried out for each identified event that could trigger an emergency shutdown. Some triggering events are incidents that activate the safety interlock system, operator action after a loss of contain-

CAUSE & EFFECT (SAFE) CHART

UNIT						
DATE						
REV	DATE	BY	CHK.	DESCRIPTION		

EFFECTS (top rows):

TAG NO	DESCRIPTION	Action
BALL-4124	PILOT FLAME OUT	Start
—	IGNITION SEQUENCE	Ann
LAH-4022	GLYCOL RECVRY SEP HI LVL	Ann
LAHH-4082	COND. SEP H H LEVEL	Ann
LAHH-4103	LIQUID SEAL H H LVL	Ann
LAL-4104	LIQUID SEAL LOW LVL	Ann
LALL-4042	GLYCL RECVRY SEP L L LVL	Ann
LALL-4046	COND SEP LO LO LVL	Ann
UOV-404	D AMP CONTROL ROOM	Close
P-404	GLYCOL RECVRY PUMP	Run
P-408	OXIDIZER LIQ FEED PUMP	SD
PAHH-4193	ASSIST GAS H H PRESS	Ann
PALL-4145	ASSIST GAS L L PRESS	Ann
TAHH-4145	UNIT HI HI TEMPERATURE	Ann
TAL-414	UNIT LOW TEMPERATURE	Ann
UA-4141	MALFUNCTION	Ann
UA-4142	SHUTDOWN	Ann
UA-4143	ESO	Ann
XSV-4021	GLYCOL RECVRY SEP VENT	Ann
XSV-4141	PILOT GAS SHUTOFF	Ann
XSV-4142	ASSIST GAS SHUTOFF	Ann
XSV-4143	ASSIST GAS SHUTOFF	Open
XSV-4144	ASSIST GAS VENT ATM	Close
XSV-4145	LIQ. SEAL O/H ESO	Close
XSV-4146	BLOWER	Close

CAUSE (columns):

TAG NO	DESCRIPTION
BSLL-4142	V-414 BOILER PILOT FLAME OUT
HS-PLT ESO	E-400 ESO FROM CONTROL ROOM
HS-4142	E-400 ENERGIZE ESO
LSH-4022	V-402 GLYCOL RECOVERY SEPARATOR HI LEVEL
LSHH-4082	V-404 CONDENSATE SEPARATOR H H LEVEL
LSHH-4103	V-410 LIQUID SEAL H H LEVEL
LSL-4104	V-410 LIQUID SEAL LOW LEVEL
LSLL-4244	V-402 GLYCOL RECOVERY SEPARATOR L L LEVEL
LSLL-4084	V-402 CONDENSATE SEPARATOR L L LEVEL
PSHH-4193	V-414 OXIDIZER ASSIT GAS H H PRESS
PSLL-4145	V-410 OXIDIZER ASSIT GAS L L PRESS
TSHH-4145	V-410 OXIDIZER H H TEMPERATURE
TSL-4144	V-410 OXIDIZER FIRE BOW LOW TEMPERATURE
ZSC-4142	V-X14 ASSIST GAS BLOCK VALVE CLOSED
ZSC-4143	V-X14 ASSIST GAS BLOCK VALVE CLOSED
ZSC-4145	V-X14 ASSIST INLET GAS VALVE CLOSED
ZSC-4148	V-X14 GAS TEMP. CONTROL VALVE CLOSED

Legend:
△ ANNUNCIATE
X PERMISSIVE
● SHUTDOWN

FIGURE 9-6. Example cause and effect diagram for a safety interlock system.

ment, equipment failure, severe process upset conditions, threatening weather conditions, and threatening conditions from adjacent process units. To refresh operator training in noncomputer-controlled facilities, pulldown charts are developed for the shutdown sequences for each possible triggering event. In computer-controlled facilities, these same individual shutdown sequences are available as special retrievable CRT panels. In the preplanning for emergency shutdown, each identified trigger event should be detectable, either by manual observation or by reliable safety-protective instrumentation. In the specific case of critical equipment failure, a process-safety review should consider the use of redundant or standby equipment to avoid the need for a sudden shutdown. Alternate strategies for isolating equipment should also be considered in the preplanning activities.

Since the safety interlock must be available when required, the preplanning activities should consider contingency situations such as the unexpected inoperability of equipment during the safety interlock sequence. Timing constraints preclude the use of management of change procedures to evaluate the effect on process safety of alternate shutdown strategies during safety interlock action.

Special conditions such as severe weather can initiate a shutdown. Many chemical plants and refineries located in coastal areas have shutdown plans that are activated when hurricanes threaten. Off-shore oil and gas production facilities also have hurricane shutdown procedures. Other weather or natural events such as earthquake, floods, severe storms, or extended freezing conditions can initiate chemical process shutdowns. Shutdowns may be caused by a lack of electric power or fuel. Sometimes these power or fuel shortages are related to weather conditions, or they may be due to equipment failures or capacity problems outside the control of the chemical facility. Accidents in adjacent facilities can also force shutdowns. The failure of a carbon monoxide sphere in the Houston ship channel area caused several adjacent chemical facilities to initiate shutdown.

9.3.2. Shutdown Sequences

Sudden shutdown occurs when a safe limit of an operating parameter is exceeded (i.e., an excessively hot pump bearing or too low a liquid level in a reactor vessel). A manually activated safety interlock shutdown can be initiated by an operator pushbutton on a local control panel or in the process unit control room in response to an alarm.

Many shutdowns are caused by circumstances not directly related to the process unit or the conditions of operation. Shutdowns have been traced to minor mistakes such as incorrectly labeled controls, calibration errors in measuring instruments, corroded electrical contacts, or the unauthorized inactivation of controls.

Traditionally, a safety system is designed to be energized in its normal state, which means that the safety interlock will activate when it is deenergized. However, some safety interlock systems are designed to activate when they are energized, thereby requiring additional safety considerations. For example, a large motor operated valve on a pipeline may not have a spring-activated closing, because of the destructive potential of hydraulic transients. An uninterruptible power source (UPS) or similar high-reliability power source should be available to close the motor operated valve on demand.

Once the process has been brought to a nonoperating but stable state, a decision should be made to hold the process in that state until the shutdown initiating event is resolved (and any necessary repairs can be made) to include inventory reduction, isolation, and decontamination. The time needed for repairs, safe holding times until repairs can be made, the economic implications of holding versus shutdown, and the need to make repairs to areas other than the cause of the shutdown should all be considered.

9.3.3. Safety Interlock Failures

Complex processes depend strongly on "engineered safety" and include many interlocks, permissives, relief or bypass schemes, and other protective devices to safeguard against accidents. The system is SAFE only if the protective devices operate when required and as designed. It is possible for protective devices to fail on demand, allowing an accident to occur.

To evaluate the safety implications of modern computer safety interlock systems, some concepts used in the logic construction of these units should be thoroughly understood.

Permissives allow the next step in an operational sequence to take place, and ensure that critical equipment components are either operational, ready to start, or in the proper position (e.g., the suction valve is open on the fire water pump before a start attempt). It is important to examine the consequences if a permissive cannot be satisfied in a particular situation. In a recent accident, the activation of a computer room halon fire extinguishing system required a positive signal showing that all the doors were closed. Because one door latch indicator was not operating, the emergency activation system would not proceed. The shutdown logic did not consider the problem of inoperative sensors, hence a small incident escalated into a catastrophic loss.

Startup bypass enables one to temporarily bypass certain shutdown input signals which, during a startup sequence, have not reached a normal operating state (e.g., lube-oil pressure may require a few seconds to reach normal levels if the lube-oil system is pressurized by a small pump powered by the main pump's shaft). Many continuous chemical process units have been designed that bypass parts of the safety interlock system during startup. Although this practice is undesirable, these safety interlock bypasses or jumpers are not uncommon. If startup bypass jumpers are the only choice, an administrative procedure to ensure that these bypass jumpers are removed should be a vital part of the startup procedure, as detailed in Chapter 7. A periodic inspection of the safety interlock system to remove the startup jumpers is also suggested.

Input bypass allows the individual disabling of any input that may be faulty by an installed jumper wire, switch, or software sequence. It does so without compromising the rest of the safety interlock system and lets the critical equipment continue operating.

A subtle point about input bypass switches is that they are the equivalent of an installed jumper wire. If the bypass switches were not there, a maintenance person, under pressure to fix a faulty safety interlock input switch causing "nuisance" shutdowns, might place a jumper wire around the input switch (intending to return later, repair the switch and remove the jumper wire). Often, "later" never comes because of the continuing pressure of higher-priority work, and it is not hard to imagine a situation where critical equipment would have many of its protective shutdown–input switches bypassed by a host of jumper wires. Because these jumper wires are essentially invisible to even an interested observer, operations staff may never become aware of them.

The jumper is installed from a necessity to keep critical equipment on-line, yet when equipment protection is really needed, it is not there and significant equipment damage can result. Because of this real danger, many owners of process plants install safety interlock input-bypass switches with indicating lights. Bypassed inputs are thus very visible to a plant inspector on tour, and faulty switches should be regularly inspected and quickly fixed.

Output bypass disables all shutdown system logic, allowing maintenance to be done on the safety interlock system logic itself without having to shut down the critical equipment. Careful management procedures should be maintained to keep the output bypass from causing future safety interlock unavailability problems. When an output bypass is used in a maintenance procedure while the process is operational, operations staff should be stationed in locations where they can begin manual shutdown. Numerous examples in the chemical process safety literature cite the unavailability of fire water pumps in emergencies because the control system was bypassed for maintenance or testing.

9.3.4. Investigation of Sudden Shutdowns

Before production is resumed after a sudden shutdown, an investigation should be performed to pinpoint the cause of the shutdown. The objective of investigating the emergency shutdown is to: (a) identify the root causes, (b) define recommendations for preventing the recurrence of a similar emergency shutdown, and (c) ensure that action is taken on the recommendations. The causes of the emergency shutdown and the findings of the subsequent investigation can then be used to consider preventive maintenance, process modifications, control changes, redundant equipment, alternate utilities, or changes in operating procedures that could help avoid the event or events that lead to an emergency shutdown.

9.4. EMERGENCY RESPONSE

A special case of the emergency shutdown is characterized by the potential or actual unplanned release of chemicals into the environment because of containment system failure. In these situations, the emergency shutdown described in the previous section should be followed by a series of additional actions, or emergency response activities.[11,12] The first- and second-line supervisors have defined roles in the execution of the emergency response plan.

9.5. SUMMARY

Since shutdown is not a routine operation, the potential for process safety problems is greater during this time than during normal operations. Accidents can occur because of the unfamiliarity of operations personnel with the shutdown sequence. During the nonoperational period, accidents may occur because the manufacturing staff is under pressure to return to operational status. Normal shutdowns and extended or mothball shutdowns are planned events with a schedule that can be distributed to all participants. The sudden or emergency shutdown is an unscheduled event that depends on the preplanning and training of operations staff and specialists in emergency response. The operations staff is responsible for performing the shutdown sequence; however, maintenance staff participation is required during all the shutdown phases. Responsibility for equipment or facilities may shift from operations to maintenance or to other groups, such as the fire brigade, specialist contractors, or outside authorities.

9.6. REFERENCES

1. *Large Property Damage Losses in the Hydrocarbon–Chemical Industries.* Marsh & McLennan, 14th Edition, New York, NY 1992.
2. *Hydrocarbon Leak and Ignition Data Base.* E&P Forum, Report 11.4/180, London, May 1992.
3. *The Flixborough Cyclohexane Disaster.* Her Majesty's Stationery Office, London, 1975.
4. Pearson, J. and Brazendale, J. "Computer Control of Chemical Plant—Design and Assessment Framework." *Preventing Major Chemical Accidents and Related Process Accidents,* IChemE Symposium Series No. 110, 1988.
5. Shaw, J.A. "Human Factors Aspects of Advanced Control." *Plant/Operations Progress,* Vol.4, No. 2, April, 1985, pp. 111–115.
6. Zodeh, O.M. and Sikora, D.S. "Self–Checking Safety Interlock System." *IEEE Transactions Industry Applications,* Vol. 25, No. 5, September/October 1989, pp. 29–33.
7. McMillan, G.K. "Can You Say Process Interlocks?" *InTech Journal,* May, 1987, p.32.

8. Guide for the Development of State and Local Emergency Operation Plans (CPG 1–8). Federal Emergency Management Agency, October 1985.
9. U.S. Department of Transportation. Hazardous Materials Emergency Response Guidebook (DOT-P-5800.5), 1990.
10. Chemical Manufacturer's Association. Community Awareness and Emergency Response Program Handbook, 1986.
11. 29 CFR Part 1910.120—Occupational Safety and Health Standards, Section 1910.120, Hazardous Waste Operations and Emergency Response, 54 FR 9294, March 6, 1989.
12. 29 CFR Part 1910.138—Occupational Safety and Health Standards, Section 1910.138(a), Emergency Action Plan.

Additional References

29 CFR 1910.165—Occupational Safety and Health Standards, Section 1910.165, Employee Alarm Systems.
Johnson E., "Reliability Assessment of Computer Systems Used in the Control of Chemical Plant." *Preventing Major Chemical Accidents and Related Process Accidents*, IChemE Symposium Series No. 110, 1988.

10

DECOMMISSIONING AND DEMOLITION

Process safety continues to play an important role in the final life cycles of a processing plant, that is, in decommissioning and demolition (D&D). Although the very technical issues of D&D are usually handled by representatives from the engineering and health, safety and environmental organizations, plenty of opportunities remain for the operations and maintenance departments to contribute meaningfully to this effort.

The decommissioning and demolition cycles may take several forms, depending on the actual events and activities at a given site. In the simplest case, a decision may be made to demolish a unit that has been mothballed, as described in Section 9.2, Extended or Mothball Shutdown. In the most extreme case, a decision may be made to shut down and demolish an operating unit or plant. Sometimes, the equipment may be salvaged for use in other processes. Other circumstances may fall in between these events. Whatever the actual situation, process safety must be properly managed to prevent an incident. Careful planning is a prerequisite to D&D and an operations representative as well as a maintenance representative should be on the planning team.

When process equipment is no longer needed for the current business plan, immediate decommissioning should be considered. Decommissioning is defined as removing the plant or unit from service. After that, it can be cleaned, maintained, and recommissioned (put back into service). It can be mothballed for later recommissioning; alternatively, the plant or equipment can be immediately dismantled or even maintained so that it can be safely demolished at a later date. Unused equipment is a potential hazard if not properly isolated from the rest of the process or plant.[1]

10.1. DECOMMISSIONING/DEMOLITION PLAN

Formal plans are needed to properly conduct decommissioning and demolition tasks. The operations and maintenance departments should participate in developing these plans, particularly decommissioning procedures. There are different combinations of decommissioning and demolition activities, each requiring its own action plan. The two basic combinations are:

- Decommissioning and immediate demolition
- Decommissioning but hold for future demolition

The decommissioning plan is different for each of these combinations. If there is no intent to refurbish the process unit, or if demolition will not be carried out immediately, the decommissioning plan should contain a periodic monitoring program for the decommissioned unit, typically conducted by the operations and maintenance departments, to ensure the continuing safe condition of the decommissioned unit.

Special provisions for holding the process equipment in a mothballed or dismantled state must be detailed and include absolute assurance that materials and energy will not enter the unit. Typically, these provisions involve completely disconnecting the piping and installation of blind flanges. In the case of electrical equipment, this includes removing fuses and breakers in addition to installing locks and tags. Special provisions for preventing any adverse effect on ongoing operations during demolition must be detailed. The decommissioning/demolition plan should provide for periodic surveys or audits to ensure that a safe storage condition is maintained. The process and utility piping systems should be reviewed to ensure that connections to utilities, drains, vent headers, and flares are completely separated from the decommissioned process unit. The removal of piping sections should be considered; closing the valves should not be relied on.

Once a firm directive to proceed with demolition has been obtained, a plan outlining the main sequence of events should be prepared. When site operations are to cease completely, it usually is necessary to consider early on the composition of a management team to supervise the demolition, since the consequences of a total shutdown may be the progressive loss of personnel as the number of employees on site is reduced. It is very important that a nucleus of operations and maintenance personnel be retained to ensure that important local knowledge is not lost and to control the issuance of permits to work and other functions. These personnel should work closely with the planning team and demolition crew to develop the demolition plan and determine the demolition sequence.

Tables 10-1 and 10-2 show examples of the work that must be done before demolition and the sequence of demolition. Once the demolition begins, technical or commercial pressures may obligate the demolition contractor to change the original plan of work. The terms of the demolition contract should

---

TABLE 10-1
Example Demolition Plan Work Prior to Demolition

1. All asbestos insulation will be cleared from the plant.

2. Surface drains will be plugged with concrete.

3. Connecting lines carrying services such as high-pressure steam and water will be isolated and disconnected.

4. All electrical supplies to the plant will be cut, except for supplies needed for demolition activity or for maintaining fire protection and emergency response capability.

5. All lubricating oils from pumps and motors will be drained and disposed of.

6. All work to reduce chemical flammability and toxicity hazards will be complete and recorded before demolition begins.

7. The plant will be cleared of all loose material and rubbish.

8. A process hazard analysis will be conducted to identify potential hazards in decommissioning and demolition activities.

9. Environmental issues, such as waste handling, waste disposal and excess flare or venting requirements will be addressed.

10. All personnel participating in the demolition will be briefed about the demolition plan, their responsibilities, the hazards that may be encountered, and safety procedures.

TABLE 10-2
Example Demolition Plan Sequence of Demolition Activities

1. Carry out the demolition in the general direction of south to north of the demolition site.

2. As they become accessible, remove equipment and components that may be retained for future use.

3. Remove XYZ storage tank and plug off feed lines. This underground tank should be filled with water before demolition work and removal.

4. Remove all pipework and valves, employing non-sparking wrenches and cold cutting techniques until all vessels, tanks, pumps, etc, are isolated from pipework.

5. Remove all pumps and motors.

6. Remove smaller vessels and heat exchangers and progressively remove steel work supporting structures.

7. Remove quench towers and compressors.

8. Remove larger tanks.

9. Remove high columns using cranes. If the flare stacks are included in the demolition, cranes should also be used for this work.

10. Remove safety and fire fighting systems last.

11. Clear the demolition site of all debris when demolition work has been completed.

make it clear that it is the contractor's responsibility to get any changes to the original plan approved by appropriate plant management. The operations and maintenance departments should participate in reviewing the proposed plan changes. During the course of the job, operations and maintenance personnel should be mindful of any contractor activities that appear to be in violation of the contract requirements.

The demolition plan should include provisions for the contractor to use the existing plant's utilities and safety equipment. Operating personnel should carefully review these provisions to ensure their workability and to control responsibility. An example guideline for safely dismantling and demolishing process units is included as Appendix B. Many details and questions must be supplied and answered by operations and maintenance staff.

10.2. OPERATION AND MAINTENANCE ROLES

After the process has been shut down, operations and maintenance personnel continue to have important roles and responsibilities in decommissioning and demolition activities. The decommissioned process will be near to or among other operating processes, and the proximity of decommissioning and dismantling activities will require some level of attention from operations and maintenance personnel. Operations and maintenance staff may be needed to carry out special decommissioning isolation, purging, or cleaning actions. Process safety incidents, such as loss of containment, excessive pressure buildup, or human error may occur if the procedures are inadequate. Long-term monitoring of the decommissioned process units may also be the responsibility of the operations and maintenance departments.

The decommissioning and demolition of a plant usually consists of three distinct steps: decommissioning, demolition preparation, and demolition. The decommissioning/demolition plan describes the activities associated with each of these phases. Engineering and environmental personnel, who receive appropriate input from the operations and maintenance departments, may be responsible for developing the decommissioning/demolition plans. In some cases, operations personnel may develop the decommissioning procedures, including decontamination. The operating history of a unit may play an important role in the development of these procedures. Knowledge of areas that may have plugged previously, areas that may be difficult to drain, areas of high corrosion, areas that may contain pyrophoric material, etc. can be incorporated into the procedures so that mitigative measures can be planned. Usually, only operating and maintenance staff will have this kind of detailed knowledge.

Once the decommissioning/demolition plan has been developed, demolition may be handled in house by the maintenance department or by specialty contractors. In either case, if demolition of the unit will not affect other process units, operations department participation in the demolition process will be less. However, a knowledgeable operations person may be needed to authorize certain aspects of the demolition work, depending on any remaining process hazards that may be present. The maintenance department may continue to play a key role in managing and controlling the demolition process, even when other process units are not affected. If the unit to be demolished is part of a larger process facility, the operations department will continue to help manage and control the work permitting process, and the maintenance department will provide close support. Close coordination of the operations, maintenance, technical groups, and demolition contractors is very important to avoiding hazardous situations that could lead to catastrophic accidents. In addition, close coordination ensures that practical plans will be developed, particularly for the part that the operations and maintenance departments must perform in the decommissioning/demolition process. The lines of communication among operations and maintenance personnel, the technical groups, and the specialty contractors must be developed before the demolition work proceeds. Operations and maintenance supervisors should monitor this communication to ensure it is effective and specific.

Lack of communication and coordination between the demolition crew and the operations personnel may be the root cause of serious process safety incidents, as illustrated by the following example. A tank to be demolished had a layer of crude oil sludge. To facilitate entry into the tank, two 5-feet square holes were needed in the north and south sides of the tank shell. The operations supervisor was contacted and a work permit was obtained. The work permit authorized using an angled grinder with a fire-proof blanket laid down and with water sprayed on the inner plates while cutting was taking place. Later, the demolition contractor decided to cut a hole in the floating roof at one of the low points to facilitate hot washing of the crude oil sludge. He mistakenly assumed that the permit for cutting holes in the tank walls was valid for cutting the hole in the floating roof. When cutting started, the sparks ignited the oil and caused a fire that resulted in injuries to several workers.

Three lessons can be learned from this incident. First, hot work should not be conducted without communicating and coordinating with personnel knowledgeable about the contents and characteristics of the hazardous material. Second, the scope of a work permit should be very clear and specific. The work permit should have clearly stated it was valid only for the first two holes and identified their location on a sketch of the tank. Third, since the work was known to have the potential to cause a fire, an operations or maintenance person should have been assigned as a "fire-watch" to observe any deviations from the work permit.

10.3. DECOMMISSIONING PROCEDURES

The principal activities involved in decommissioning are removing the chemical inventory, isolation, purging, and cleaning. Removing the chemical inventory is normally the first step in decommissioning a process unit. Most of the liquid or gas inventory can be removed with available process units and drain systems. It may be necessary to isolate certain process streams. The removal of chemical solids such as catalysts or special packing materials may require additional specialized equipment. The removal of residual chemicals, sometimes referred to as decontamination, requires careful consideration. Chemicals in the vapor state are often removed by purging; liquid (or solid) chemicals may have to be washed with water or solvents.

After purging and cleaning is complete, the entire process, including utility and electrical systems, is then physically isolated. Operations and maintenance personnel should carefully examine the actual procedures and sequence of activities to ensure that the plan can be executed safely and that no unrecognized process hazards remain when the decommissioning is complete.

Purging,[2,3,4] which consists of replacing one fluid by another in an enclosed space, is accomplished by two distinct actions — displacement or dilution. Displacement is usually selected for purging; however, equipment with an irregular shape or piping with recycle loops may have to be diluted to remove the process components. If purging is carried out with inert gases, purging by dilution can best be accomplished by pressuring and venting the system until the lower flammable or lower toxicity limit is reached. For example, in ethane/ethylene separation plants, purging is usually accomplished by pressuring up with heated nitrogen and then venting. The procedure is repeated a number of times until the concentration of hydrocarbons is at a nonhazardous level. Three to four cycles are usually sufficient when air can be introduced into the system. A number of incidents have been caused by improper purging procedures and a failure to test for hazardous gases.

An incident occurred during the purging of hydrogen from a naphthalene hydrodealkylation system. After every cycle of pressuring and depressuring, an explosimeter was used to check the concentration of hydrogen. Unfortunately, the explosimeter was calibrated for methane. Thus, when plant personnel were sure that the purging was sufficient, in actuality, the concentration of hydrogen was still within the flammable limits. The mixture was released and encountered an ignition source, causing a fire and explosion. The incident resulted in serious injuries to a number of workers. This example demonstrates the importance of proper procedures, instrumentation, and training for personnel involved in the purging process.

When the contents of a process unit and associated piping must be changed from a flammable material to air, the formation of flammable mixtures may be avoided by using a double purging procedure. Double purging consists of interposing an inert substance between the two media being interchanged. For example, when a piece of equipment is taken out of service for decommissioning, the combustible gas contents of the equipment may be first replaced by an inert gas, such as nitrogen, steam, or carbon dioxide. The inert gas is later replaced by air.

Note that purging and dilution activities may involve the flow of fluids (types, volumes, directions) for which the original process was not designed. Operations personnel are well suited to examine such flows for potential process hazards. Likewise, it is especially important to properly test the atmosphere inside the equipment for flammability, toxicity, and oxygen content before beginning demolition operations. Operations personnel also have an important role in testing.

Many serious injuries and deaths have occurred during demolition activities from asphyxiation. It cannot be overstated how important it is to test the atmosphere of equipment immediately before starting demolition even though the equipment may have been properly decommissioned. In one case, a tank that had contained only water was being prepared for demolition.[5] Since it contained only water, the usual tests of the tank atmosphere were not carried out. Three persons entered the tank and were overcome. Two recovered, but one died. The atmosphere inside the tank was afterwards found to be deficient in oxygen. The investigators believe that rust formation used up some of the oxygen after the tank had been drained. This incident demonstrates the need for a careful evaluation and enforcement of procedures, even for apparently safe equipment, as well as for potentially hazardous equipment, before starting demolition activities. Severe corrosion can also liberate hydrogen, which could create a flammable atmosphere inside the process equipment.

The importance of purging equipment in hazardous materials service before its dismantling and demolition is well recognized; however, the procedures are sometimes not implemented correctly. In one case, a reactor was isolated and prepared for demolition. It was purged and welders were allowed to enter the reactor to start the dismantling. After doing some work inside the reactor, the welders worked for a period on the outside of the reactor. Later, they had to reenter the reactor for additional dismantling and they did so without carrying out any atmosphere tests. When welding started, an explosion occurred, killing two of the workers and seriously injuring the other. A later investigation determined that between the two entries, flammable gas had leaked into the vessel through a leaking tube.

At least three lessons can be learned from this incident. First, the demolition procedure should have followed regular permit-to-work procedures (e.g., confined space entry, hot work permit, etc.), which would have required continuous or periodic monitoring of the space inside the reactor. Second, it should never be assumed that a unit or piece of equipment that has been decommissioned and prepared for demolition ceases to be a hazard. Third, proper application of lock out-tag out procedures would have prevented flammable gas from entering the vessel.

Selection of the inert medium is important. Although steam is easily available, it may not be the best medium for purging. For one, steam purging may damage equipment. An incident occurred during the purging of a flare line in a hydrogen purification plant, where nitrogen was usually used for purging. However, on one occasion when nitrogen was not available, low-pressure steam was used. The aluminum flare line expanded, causing the pipe shoes to move off the pipe support, as shown in Figure 10-1. This allowed the pipe to drop directly onto the support. When the line cooled, it contracted and the shoes caught on the support and were ripped off the pipe, rupturing the line. In addition to causing a significant amount of property damage, the ruptured line released flammable gas that a nearby ignition source ignited, resulting in severe injuries to two workers. This incident demonstrates the need to properly consider the inert medium used, including the potential adverse effects, before purging is started. Also, management of change procedures should have been followed before deciding to change the inert medium from nitrogen to low-pressure steam. Table 10-3 shows other examples of equipment damaged by steam purging.

The inadequate cleaning of process units can lead to future process safety problems. For example, an acrylonitrile manufacturing unit was to be mothballed for several months because of low product demand. The absorber section was washed with acid to remove accumulated polymer. After acid washing, the procedure for flushing and neutralizing the acid was not followed correctly, leaving acidic conditions in the absorber column and some of the heat exchangers. The acidic condition was not monitored over the several months when the unit was mothballed. When the unit was started up,

---

TABLE 10-3
Examples of Equipment Damaged by Steam Purging

Equipment	Reason for damage
Low-temperature piping and equipment	Design did not allow for expansion caused by high-temperature process fluid
Piping with bitumastic or asphaltic coatings	Damage to coating not rated for steam temperature
Cast iron equipment	Unequal heating caused stress and subsequent fracture
Close tolerance equipment	Thermal expansion from steam temperature caused damage to bearings, shaft, etc.
Equipment with large heat capacitance or surface area	Design did not allow for handling large quantities of steam condensate
Tanks containing flammable mixtures	Static discharge and explosion with water droplets in the steam

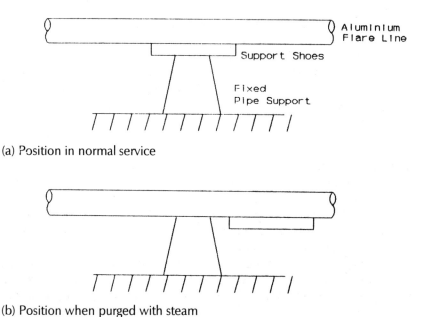

(a) Position in normal service

(b) Position when purged with steam

FIGURE 10-1. Illustration of example showing the importance of proper selection of inert medium.

several of the trays were dislodged and fell because corrosion had weakened the assembly bolts and hold down clips. In addition, two heat exchangers developed tube leaks from corrosion. This example, although it did not lead to a loss of containment, demonstrates the need to adequately clean process units during decommissioning. Cleaning is particularly important if plans call for restarting the unit in the future or for salvaging the equipment. Likewise, cleaning is important if the process material is toxic and the equipment must be disposed of. Operations and maintenance personnel may have specific knowledge about the completeness of a cleaning operation.

Heat exchanger cleaning procedures may require special techniques to clean plugged tubes within the tube bundle. If the plugged tubes are not clean, flammable gas pockets in the plugged tubes may explode during dismantling or cutting operations. For example, the heat exchanger tube bundle in an acrylonitrile plant was pulled for dismantling. The acrylonitrile production process used a small quantity of hydrogen cyanide. Also, some of the tubes were plugged with polymer. When the demolition crew was cutting into the tubes, the cyanide and water inside the plugged tubes was released and injured the workers. Again, operations and maintenance personnel are probably in the best position to understand the specific details of cleaning equipment. This knowledge should be built into the demolition plan through the review process.

10.4. MAINTENANCE OF DECOMMISSIONED STATUS

Clear labeling or tagging of decommissioned process units is necessary to avoid activities that may compromise the mothballed state of the unit and thus cause an incident. The initiating event for many of these incidents is usually the opening of a valve or a line or the removal of a piece of equipment for use as a spare part.

Using equipment from decommissioned units for spare parts can also cause serious incidents. In a specialized proprietary process, a rubber additive material was being produced in a reactor. During a graveyard shift, one of the reactor bottom pumps started malfunctioning. The maintenance technician decided to replace the pump and have it repaired later. The warehouse inventory did not have a replacement pump. The technician remembered that he had seen a similar pump in a nearby mothballed unit. On checking the pump, he found the specifications to be satisfactory. The pump was taken out and installed in the operating process. The mechanical integrity of the pump was not verified. On restarting the operation, the pump failed and released a large quantity of toxic material, killing one worker and severely injuring two others. One lesson that can be learned from this incident is the importance of using management of change procedures. In addition, strict procedures for ensuring quality assurance and verifying mechanical integrity should be used before mothballed equipment is used as spare parts.

Equipment should be monitored for corrosion if portions of the decommissioned unit may be reused. Similarly, periodic monitoring for abnormal pressure or temperature conditions may be appropriate. This is particularly important for units that have not been cleaned thoroughly and that contain sludges that can liberate gases or otherwise react when out of service.

Isolating equipment requires additional measures beyond closing the valves, as noted earlier in this chapter, as well as in the section on extended shutdowns in Chapter 9. Valves can develop leaks over a period of time because of temperature cycling effects or corrosion. Thin blinds used for slight positive pressure seal requirements may not be strong enough over extended periods. Water seals are not reliable over a long period because of the potential for water to evaporate. Process units that must be held under positive pressure to avoid atmospheric contamination should be checked and tested periodically to ensure the integrity of the seal. Monitoring the rise and fall of pressure over time may also be required. Holding purge or blanket gases for an extended period requires a considerable monitoring effort by the operations and maintenance departments.

The evolution of gas from insulation may be observed in certain decommissioned process equipment, requiring the periodic monitoring of pressure and gas composition. This usually happens in insulated vessels used for LNG service. An accident occurred on Staten Island in 1974 during the demolition of an LNG tank. The tank had been purged and tested to be free of hazardous material; however, when workers started the dismantling, using burners, the LNG adsorbed in the insulation started to desorb and was ignited. A major fire resulted, which spread to other equipment very rapidly and killed 40 people. Had a process hazards analysis been performed, it might have identified the potential for gas to evolve from the insulation.

Tarry deposits or aromatic tar sludges on the inside of vessels and tanks in certain processes can ignite during demolition. Special cleaning or water-filled conditions may be required. For example, in one case, a tank with gummy deposits on the walls and roof had to be demolished.[5] The deposit was unaffected by steam but gave off vapor when a burner's torch was applied on the outside. The vapor exploded, killing six firemen who were on the roof at the time. Since it is not always possible to completely remove such sludges or deposits, special procedures should have been used during the demolition. One such procedure is filling the tank with inert gas or fire-fighting foam generated with inert gas.

As discussed in Section 10.1, an audit should be performed to verify the proper implementation of decommissioning/demolition plans. During the audit, the auditor should verify that procedures were in place that specified the general requirements for isolating and cleaning the equipment to be taken out of service and that these procedures were adhered to. Likewise, the auditor should evaluate how the decommissioning was handled, including adherence to any special cleaning procedures or special disposal procedures, such as asbestos disposal. For mothballed equipment, the audit should be repeated periodically to ensure that the equipment is maintained in the state described in the decommissioning/demolition plans.

10.5. DEMOLITION CONCERNS

Specialty contractors are typically used for demolition tasks, although some companies may use their own maintenance departments. Operations and maintenance departments need to coordinate with the demolition team, particularly if the demolition is within or adjacent to an operating unit. Demolition operations can create more serious problems than construction operations if they are not properly controlled. Crane accidents and dropped-object accidents affecting adjacent operating chemical process units have been widely reported in the literature.

Many accidents occurring during demolition work are the result of improper procedures, the lack of enforcement of safety rules, and general carelessness. Some governmental regulations require predemolition surveys; for example, OSHA requires[1] a predemolition survey of buildings and structures.

---

1 In the United States, OSHA's demolition standard [*29 CFR 1926.850(a)*] requires that an

The purpose of the survey is to determine the condition of the framing, floors, and walls so that measures can be taken, if necessary, to prevent the premature collapse of the structure. However, even if surveys are made, careless implementation of the demolition plan may still lead to potentially hazardous incidents.

For example, during the demolition of a refinery, a propane cylinder split open and caught fire when a section of column rolled and a piece of pipework attached to it struck the cylinder. The accident seriously injured several workers. One of the lessons learned from this incident is that demolition work should be carried out in a safe and controlled way through proper work permits. Also, major demolition work is often very complicated, requiring detailed preplanning and close supervision. In general, flammable material should be removed from the work site before demolition begins.

Demolition operations should include precautions to ensure that the safety-protective or emergency response equipment is maintained as long as practicable. The fire protection equipment should be maintained in proper working condition until all demolition activities have been completed. Utilities needed to maintain fire protection and emergency response capabilities may have to be relocated or protected.

The removal of equipment from the demolition site, particularly items that may be going to scrap, needs to be considered carefully. The demolition plan needs to establish a procedure that calls for the inspection of cleaned equipment to ensure that all hazardous material has been removed. The plan should specify who authorizes the removal of the equipment and certifies that it is free of hazardous material.

The demolition plan should also address paints and coatings that may emit toxic fumes during flame cutting. Medical monitoring may be advantageous and, if so, it should be stipulated. Other potentially toxic or noxious operations should be covered. These steps may be beyond the knowledge of the operations and maintenance personnel, but they can still perform a valuable service by understanding the requirements of the demolition plan and being on the lookout for any violations.

10.6. SUMMARY

Although decommissioning and demolition activities are specialized and occur occasionally, the operations and maintenance staff still have an important process role to fulfill. Their in-depth knowledge of a unit or plant, gathered during the many years of the operating life cycle, are invaluable to accomplishing the broad aspects of decommissioning and demolition. This knowledge is needed to:

engineering survey of the structure be conducted by a competent person.

- Prepare a decommissioning and demolition plan that is both safe and workable
- Ensure that the plan outlines steps to effectively remove all hazardous process materials
- Carry out the decommissioning plan, including the purging and decontamination steps
- Test effluent streams to ensure that flammability, toxicity, and oxygen limits have been achieved during decommissioning
- Monitor the continued safe status of a decommissioned process if demolition is delayed
- Test the atmosphere of equipment and systems immediately before demolition starts
- Be alert to any deviations from the plan during actual demolition operations
- Be alert to any management of change issues
- Take part in communications with a demolition contractor, particularly regarding operations in or near operating units
- Issue proper, thoughtful work permits
- Serve as a "fire-watch"
- Ensure that equipment isolation steps are not violated
- Observe the work site each day for potential hazards involving a lack of cleanliness or obstructions
- Perhaps serve as the agent to approve the removal of process equipment from the demolition site
- Assign transportation routes for moving demolition construction equipment or removing process equipment
- Be alert to implement the emergency preparedness plan in the unlikely event of an incident in the demolition area

10.7. REFERENCES

1. Barber, D. H. *Safety in Petroleum Refinery Operations.* Proceedings of the Ninth World Petroleum Congress, Vol. 6, "Conservation and Safety." Applied Science Publishers, Ltd., London, 1975.
2. U.S. Coast Guard. *Manual for Safe Handling of Inflammable and Combustible Liquids.* Document #CG174, Washington, D.C. 1951.
3. National Fire Protection Association. *Flammable Liquids Code.* NFPA 30, Boston, 1991.
4. American Gas Association. *Purging Principles and Practice.* New York, 1954.
5. Kletz, T. A. *What Went Wrong? Case Histories of Process Plant Disasters.* Gulf Publishing Company, Houston, 1988.

APPENDIX

A

SUMMARY OF THE PROCESS SAFETY MANAGEMENT RULE[1] PROMULGATED BY THE OCCUPATIONAL SAFETY AND HEALTH ADMINISTRATION, UNITED STATES DEPARTMENT OF LABOR

Background

The fourteen elements of the Process Safety Management (PSM) rule (29 CFR 1910.119) were published in the *Federal Register* on Monday, February 24, 1992. The objective of the rule is to prevent or minimize the consequences of catastrophic releases of toxic, reactive, flammable, or explosive chemicals. The rule will accomplish its goal by requiring a comprehensive management program: a holistic approach that integrates technologies, procedures, and management practices. Process Hazards Analysis may not be performed unless complete Process Safety Information is available for the process. The compliance deadline for Process Safety Information and Process Hazards Analysis is as follows:

1 This appendix summarizes the Process Safety Management (PSM) regulation (§ *29 CFR 1910.119*) promulgated by the Occupational Safety and Health Administration, U.S. Department of Labor on February 24, 1992. Users of this document should consult the *Federal Register* for a complete understanding of the legal requirements. The American Institute of Chemical Engineers, its consultants, CCPS Subcommittee members, their employers, their employers' officers and directors, and RMT/Jones and Neuse, Inc. accept no liability for any regulatory impact that may occur at any facility as a result of differences between this summary and the PSM regulation given in the *Federal Register*.

- At least 25% must be completed by May 26, 1994.
- At least 50% must be completed by May 26, 1995.
- At least 75% must be completed by May 26, 1996.
- 100% must be completed by May 26, 1997.

In addition to the twelve elements contained in the proposed rule, OSHA has added two more elements to the final rule. They are: employee participation and trade secrets. A brief summary of the various requirements of the PSM rule is given here.

Application §1910.119(a)

- Applies to processes that involve chemicals at or above threshold quantities (specified in Appendix A) and processes which involve flammable liquids or gases onsite in one location, in quantities of 10,000 pounds or more (subject to few exceptions). Hydrocarbons fuels, which may be excluded if used solely as a fuel, are included if the fuel is part of a process covered by this rule.
- Does not apply to retail facilities, oil or gas well drilling or servicing operations, or normally unoccupied remote facilities.
- Although not part of this rule, 29 CFR 1910.109 (Explosives and blasting agents) has been revised to indicate that manufacturers of explosives and pyrotechnics must meet the requirements contained in 29 CFR 1910.119.

Employee Participation §1910.119(c)

This element of the rule requires developing a written plan of action regarding employee participation; consulting with employees and their representatives on the conduct and development of other elements of process safety management required under the rule; providing to employees and their representatives access to process hazard analyses and to all other information required to be developed under this rule.

Process Safety Information §1910.119(d)

This element of the PSM rule requires employers to develop and maintain important information about the different processes involved. This information is intended to provide a foundation for identifying and understanding potential hazards involved in the process.

The Process Safety Information covers three different areas: chemicals, technology and equipment. A complete listing of the process safety information that must be compiled in these three areas is shown in Table A-1.

Process Hazards Analysis §1910.119(e)

This element of the PSM rule requires facilities to perform a process hazards analysis (PHA). The PHA must address the hazards of the process, previous hazardous incidents, engineering and administrative controls, the consequences of the failure of engineering and administrative controls, human factors, and an evaluation of effects of failure of controls on employees. This element requires that the PHA be performed by one or more of the following methods or any other equivalent method:

- What-if
- Checklist
- What-if/Checklist
- Hazard and Operability (HAZOP) studies
- Failure Modes and Effects Analysis (FMEA)
- Fault Tree Analysis

The rule suggests a performance oriented requirement with respect to the PHA so that the facility will have the flexibility to choose the type of analysis that will best address a particular process.

TABLE A-1 Process Safety Information		
Chemicals	*Technology*	*Equipment*
Toxicity	Block flow diagram or simplified process flow diagram	Materials of construction
Permissible exposure limits	Process chemistry	Piping and instrumentation diagrams
Physical data	Maximum intended inventory	Electrical classification
Reactivity data	Safe limits for process parameters	Relief system design and design basis
Corrosivity data	Consequences of deviations	Ventilation system design
Thermal and chemical stability data		Design codes and standards employed
Hazardous effects of inadvertent mixing of different materials		Material and energy balances
		Safety systems

Operating Procedures §1910.119(f)

The operating procedures must be in writing and provide clear instructions for safely operating processes; must include steps for each operating phase, operating limits, safety and health considerations and safety systems. Procedures must be readily accessible to employees, must be reviewed as often as necessary to assure they are up to date and must cover special circumstances such as lockout/tagout and confined space entry. The employer must certify annually that the operating procedures are current and accurate.

Training §1910.119(g)

The rule requires that facilities certify that employees responsible for operating the facility have successfully completed (including means to verify understanding) the required training. The training must cover specific safety and health hazards, emergency operations and safe work practices. Initial training must occur before assignment or employers may certify that employees involved in the process as of May 1992, have the required knowledge, skills and abilities to safely perform duties and responsibilities specified in the operating procedures. Refresher training must be provided at least every three years.

Contractors §1910.119(h)

The rule identifies responsibilities of the employer regarding contractors involved in maintenance, repair, turnaround, major renovation or specialty work, on or near covered processes. The PSM rule requires the employer to: consider safety records in selecting contractors; inform contractors of potential process hazards; explain the facility's emergency action plan; develop safe work practices for contractors in process areas; periodically evaluate contractor safety performance; and maintain an injury/illness log for contractors working in process areas. The rule also requires the contractor to train their employees in safe work practices and document that training, assure that employees know about potential process hazards and the employer's emergency action plan, assure that employees follow safety rules of facility, advise employer of hazards contract work itself poses or hazards identified by contract employees.

Pre-Startup Safety Review §1910.119(i)

This element of the PSM rule requires a pre-startup safety review of all new and modified facilities to confirm integrity of equipment; to assure that appropriate safety, operating, maintenance and emergency procedures are in place; and to verify that a process hazard analysis has been performed. Modified facilities for this purpose are defined as those for which the modification required a change in the process safety information.

Mechanical Integrity §1910.119(j)

This element of the PSM rule mandates written procedures, training for process maintenance employees and inspection and testing for process equipment including pressure vessels and storage tanks; piping systems; relief and vent systems and devices; emergency shutdown systems; pumps; and controls such as monitoring devices, sensors, alarms and interlocks. PSM calls for correction of equipment deficiencies and quality assurance activities to ensure new equipment and maintenance materials and spare parts are suitable for the process and properly installed.

Hot Work Permit §1910.119(k)

This element of the PSM rule mandates a permit system, as described in 29 CFR 1910.252(a), for hot work operations conducted on or near a covered process. The purpose of this element of the rule is to assure that the employer is aware of the hot work being performed, and that appropriate safety precautions have been taken prior to beginning the work.

Management of Change §1910.119(l)

This element of the rule specifies a written program to manage changes in chemicals, technology, equipment and procedures which addresses the technical basis for the change, impact of the change on safety and health, modification to operating procedures, time period necessary for the change, and authorization requirements for the change. The rule requires employers to notify and train affected employees and update process safety information and operating procedures as necessary.

Incident Investigation §1910.119(m)

This element of the rule requires employer to investigate as soon as possible (but no later than 48 hours) incidents which did result or could have resulted in catastrophic releases of covered chemicals. The rule calls for an investigation team, including at least one person knowledgeable in the process (and a contractor employee, if appropriate), to develop a written report of the incident. Employers must address and document their response to report findings and recommendations and review findings with affected employees and contractor employees. Reports must be retained for five years.

Emergency Planning and Response §1910.119(n)

This element requires employers to develop and implement an emergency action plan according to the requirements of 29 CFR 1910.38(a) and possibly an emergency response plan in accordance with 29 CFR 1910.120.

Compliance Audits §1910.119(o)

This element of the rule requires employers to certify that they have evaluated compliance with process safety requirements every three years and specifies retention of the audit report findings and the employer's response. Employer must retain the two most recent audits.

Trade Secrets §1910.119(p)

Similar to the trade secret provisions of the hazard communication rule requiring information to be available to employees from the process hazard analyses and other documents required by the rule. The rule permits employers to enter into confidentiality agreements to prevent disclosure of trade secrets.

Appendix A: List of Highly Hazardous Chemicals (Mandatory).

Contains a list of 137 toxic and reactive chemicals which present a potential for a catastrophic event at or above the threshold quantity. Provides a threshold quantity for each chemical listed.

Appendix B: Block Flow Diagram and Simplified Process Flow Diagram (Nonmandatory)

Appendix C: Compliance Guidelines and Recommendations for Process Safety Management (Nonmandatory).

Contains guidelines to assist employers in complying with the requirements.

Appendix D: Sources of Further Information (Nonmandatory).

Lists organizations and documents that provide further information on process safety management.

SUMMARY COMPARISON OF OSHA ELEMENTS WITH CCPS ELEMENTS

It is important to point out that OSHA's PSM elements as given in §29 CFR 1910.119 are similar to or completely contained in the CCPS elements. The following comparison matrix shows the relevance of OSHA elements to CCPS's chemical process safety management elements.

CCPS's Twelve Elements of Chemical Process Safety Management	Relevant Paragraphs of OSHA's PSM Rule
1. Accountability: Objectives and Goals	
2. Process Knowledge and Documentation	Process Safety Information §1910.119 (d)
3. Capital Project Review and Design Procedures (for new and existing plants, expansions, and acquisitions)	Pre-Startup Safety Review §1910.119 (i) Mechanical Integrity §1910.119 (j)
4. Process Risk Management	Process Hazard Analysis §1910.119 (e) Pre-Startup Safety Review §1910.119 (i)
5. Management of Change	Management of Change §1910.119 (l)
6. Process and Equipment Integrity	Process Hazard Analysis §1910.119 (e) Operating Procedures §1910.119 (f) Mechanical Integrity §1910.119 (j)
7. Human Factors	Process Hazard Analysis §1910.119 (e) Operating Procedures §1910.119 (f)
8. Training and Performance	Operating Procedures §1910.119 (f) Training §1910.119 (g) Pre-Startup Safety Review §1910.119 (i) Emergency Planning and Response §1910.119 (n)
9. Incident Investigation	Incident Investigation §1910.119 (m)
10. Standards, Codes, and Laws	Compliance Audits §1910.119 (o)
11. Audits and Corrective Actions	
12. Enhancement of Process Safety Knowledge	

B

EXAMPLE MANAGEMENT GUIDELINES FOR THE SAFE DISMANTLING AND DEMOLITION OF PROCESS PLANTS

1.0. INTRODUCTION

The advent of new technology and changes in the chemical and processing industry often make it necessary to dismantle and demolish redundant facilities, ranging in scope from individual units or plants to entire locations. Many potential hazards are associated with these activities, and the requirements of the company policies on health, safety and environmental have an important bearing on how they are carried out.

Whenever a unit, plant or location becomes redundant, it is company policy to ensure as far as possible that safe conditions are maintained after operations cease. Following the complete closure of operations at a location, for example, management may decide that all equipment should be dismantled and removed, structures should be demolished, and ground reclaimed where necessary to ensure that no significant hazards remain. All this must be done with due regard to health, safety and the environment.

When a location is to be sold to a third party who proposes to continue operations, total demolition may not be necessary, although parts of the plant or location may need to be dismantled. Maintaining safe conditions and the security of the location will remain the responsibility of company management, however, up to the time of handover.

Special considerations may be needed for work involving cross-country pipelines where there is likely to be a major effect on third parties outside the boundaries of a defined location.

This document discusses the health and safety issues that should be taken into account when preparing for and executing demolition operations. Checklists are presented for use before and during the course of a demolition contract. The document also emphasizes the need to exercise particular care over the choice of a competent contractor to ensure safe working conditions.

2.0. PREPARATION FOR DEMOLITION OPERATIONS

2.1. Planning

Once a firm directive to proceed with demolition has been obtained, a plan outlining the main sequence of events should be prepared. In cases where operations are to cease completely, the composition of a company management team to supervise the demolition must be considered early on, since the consequences of a total shutdown will be the progressive loss of personnel as employment on the site is reduced. Wherever possible, a nucleus of experienced supervisory staff and key operating employees should be retained to ensure that relevant knowledge is not lost and to control the issuance of safe work permits, perform atmosphere tests, and conduct other tasks that are the responsibility of on-site management. The need also to retain a small number of administrative staff during the phaseout should not be overlooked.

Many states have a legal requirement to inform the statutory safety agency(ies) when demolition is to take place. If appropriate, early contact with the agency should be arranged to ensure that this and any other federal, state, or local regulatory requirements are met.

2.2. Preparation of Equipment

Before demolition work is allowed to proceed, steps should be taken to remove or otherwise make safe potentially hazardous substances and services in the area to be demolished, as far as is possible during the plant decommissioning phase. Examples include:

2.2.1 Residual feedstocks, process materials and products in pipelines, piping and equipment, storage vessels, and drains.

2.2.2 Catalyst materials in process vessels.

2.2.3 Residual chemical stocks in treatment systems.

2.2.4 Surface deposits such as lead in tankage, pyrophoric iron, vanadium in fired heaters and boilers and tars, sludges, and chemical solids in tanks and vessels.

2.2.5 Insulation materials, particularly asbestos.

2.2.6 Electrical power supplies, water, steam, compressed air and any other services.

2.2.7 Interconnecting piping (e.g., drains, flares, or blowdown systems).

2.2.8 All sources of ionizing radiations.

Records should be kept that indicate the condition of each plant item, the date of preparation, and the person or group responsible for releasing an item or system for demolition.

Any serviceable equipment to be retained or sold should be removed from the demolition area and placed in storage once it has been made safe. It is advantageous to prepare an inventory of equipment items to enable moveable assets to be accounted for at all stages of the project.

2.3. Disposal of Materials

During the course of demolition operations, material will usually have to be removed from the location and land may have to be reclaimed. Site management should, therefore, consider the health and safety implications of any requirements for:

2.3.1 Dumping rubble or spoil material and any road traffic movements.

2.3.2 Disposing of contaminated soil and sludges.

2.3.3 Cutting up and removing plant components.

2.3.4 Burning wood and other combustibles, paying attention to the emission of toxic fumes from contaminants.

2.3.5 Draining lakes and reservoirs, pumping out sumps, tanks, and drains.

2.3.6 Filling in shafts, pits, trenches, sumps, tunnels, etc.

2.3.7 Removing or making safe underground piping, cabling, foundations, and piling. Removing superstructure and the pilings of jetties. Issues include access to and method of excavating cross-country pipelines, particularly where pipelines cross rivers, roads, railways, and utilities.

2.3.8 Landscaping or regrading the location, reclaiming land and fences for cross-country pipelines.

2.3.9 Transporting or towing heavy plant or structures to another site or to another location for breaking.

3.0 PREPARATION OF CONTRACTOR BID SPECIFICATION

3.1. Survey of Location

As stated in Section 1, the company retains responsibility for the overall safety of the location until the conclusion of demolition operations, although contractors will normally be engaged to carry out most of the work. The site should therefore be surveyed after initial preparation to establish the nature and extent of the hazards that may exist and to facilitate the assembly of an adequate bid document. During this and subsequent phases of the project, the participation of senior operating and maintenance personnel familiar with the operations will be vital, because detailed knowledge of matters such as electrical services, isolation systems, drains, methods for thorough cleaning contaminated equipment, interaction with third parties in the case of pipelines and jetties etc., is needed to avoid hazardous conditions. A process hazards analysis will probably be needed.

The survey report should cover at least the following items:

3.1.1   What plans are available of the unit, plant, location, and the surrounding area that show access roads, services to the location, the relationship of any pipelines to nearby third-party locations, and changes or alterations to the original arrangement?

3.1.2   Does the location of the plant or unit mean that during demolition, adjacent property or people in the neighborhood may be affected by noise, dirt, vibration etc., or put at risk by the collapse of structures?

3.1.3   What are the electrical zone classifications of adjacent areas where operations may continue during the demolition?

3.1.4   Is access to the unit, plant, or location safe both above ground and below ground for initial inspection?

3.1.5   Has the unit, plant, or location been adequately decommissioned or otherwise made safe? What records are available? Have tanks and systems that contained hazardous process materials been cleaned out, purged, or inerted? Will a continuing check on the tank contents for flammable or toxic materials be required during the project? Are the roofs of floating roof tanks adequately supported? What health and safety requirements will there be for entry to columns or vessels? Are lockout/tagout requirements strictly enforced?

3.1.6   Are there any equipment items that require special attention to deal with remaining toxic materials? (e.g., the presence of asbestos, lead compounds, etc.)

3.1.7   Do any areas exist where hazardous substances may have collected? Do any vessels have separate linings that may release flammable or toxic residues during demolition? Is there any contaminated soil?

3.1.8 What is under the ground? Does the site contain areas where heavy loads cannot be placed safely? Are new access roads needed? Are there any conditions that will limit excavations?

3.1.9 Are there any limiting height conditions that may restrict operations (especially overhead power lines)?

3.1.10 Can the unit, plant, or location be easily isolated from all external services? Do some services need to be retained or diverted, electricity, air, water, steam, sewage etc., and can these services be adequately identified?

3.1.11 What are the existing provisions for fire fighting and emergency medical services? Are they adequate? If not, how will adequate coverage be provided? Is adequate telephone communication to outside emergency services available? Is an emergency preparedness plan in place?

3.1.12 Do external authorities have to be notified? Who will be responsible for this?

3.1.13 Will it be necessary to protect adjacent plant or property from the effects of demolition (e.g., ground vibration, dust, missiles)?

3.1.14 Will demolition affect the integrity of adjacent plant or property? Could land slip occur? Will people have to be evacuated? Could hazardous substances leak or be emitted?

3.1.15 How will safe areas be defined or allocated for the contractor to house staff, provide eating and washing facilities, and park vehicles?

3.1.16 Could the local climate or weather adversely affect the safety of the demolition (e.g., wind, heat, cold, snow, tides, waves)?

3.1.17 How will safe areas be defined for cutting, cleaning, storing, or handling dismantled materials without posing danger to location operators or other personnel?

3.1.18 What security arrangements will be needed to protect adjacent plant or property?

3.1.19 How will the removal of commercially valuable scrap metals (nonferrous, mercury, etc.) be controlled and monitored to protect the interests of all involved?

3.1.20 Have precautions been made for potential hazards identified in the process hazards analysis?

3.2. Health, Safety, and Environmental Review

At a suitable point before issuing the invitation to bid, a formal review should take place to ensure that, as far as possible, all aspects of the work relating to health, safety, and the environment have been adequately considered and included in the bid documents.

The review team should consist of representatives from health, safety, and environmental security and from the operations and engineering departments. The team should include members with as much direct experience of and familiarity with the operation of the site as possible. Key operating and maintenance staff may play an important role.

The matters under review should include the following:

3.2.1   The scope of the work to be undertaken should be defined as precisely as possible.

3.2.2   Each item or service to be isolated should have been identified and the method of isolation defined. Double block and bleed techniques are not acceptable. All piping, ducting and cables, both above and below ground, must be taken into account.

3.2.3   All hazardous substances likely to be encountered during the demolition/dismantling phase should be identified and information gathered about the appropriate handling and disposal, methods taking into account any regulatory requirements.

   If practical, all harmful substances, particularly asbestos, should be removed under company supervision before the demolition contractor begins work.

   Some of these operations, for example, the disposal of lead compounds and sludge, may require the use of specialist contractors or agencies.

   The objective of these precautions is to place as much of the installation as possible into a clean state, free of oil, gas, chemicals, and other residues, to reduce the possibility of an incident arising from error or misunderstanding once the contractor is on site.

3.2.4   Systems and procedures for controlling hot work should be critically examined.

3.2.5   Any limiting conditions or circumstances that a contractor would need to be aware of to prepare a realistic bid should be noted (i.e., limits on hours of work, operation of the location permit system, environmental considerations.)

3.2.6   Requirements for initial medical examination of all the contractors' staff should be considered. The is particularly relevant where work that may involve the burning of lead or lead-coated materials is concerned.

4.0 CONTRACTOR SELECTION

4.1. Contractor Competence

To carry out demolition operations, it is normally necessary to employ a contractor, who may, in turn, find it necessary to use subcontractors and to lease special equipment they themselves do not own.

The overall safety of demolition will be enhanced by the choice of a suitable contracting organization. To protect the interests of the company and its employees, and in certain cases for legal reasons, individuals who select contractors should take steps to ensure that a responsible and appropriately experienced company is awarded the bid. If a trade organization that regulates dismantling and demolition contractors resides in the state concerned, site management should ensure that they include members of that body on the bid list. Also, management should carry out a precontract evaluation of the contractors who submit bids, looking at safety awareness and competence, as well as relevant experience and accident statistics. It may be possible to arrange a visit to a location where a particular contractor is working to gather first-hand experience of the methods the company used.

Corporate health, safety, and environmental have published guidelines on employing contractors, and these recommendations should be followed where appropriate. Issues that should be considered when selecting a contractor should include at least the following:

4.1.1 Is the contractor a member of a national federation of dismantling and demolition contractors or of an appropriate regulatory body? Can the contractor demonstrate a familiarity with recognized standards such as OSHA's Demolition Standard 29 CFR 1926.850?

4.1.2 What previous experience has the contractor had with similar projects? Bidders should be required to submit an outline plan of the sequence of demolition and the methods they would use if selected.

4.1.3 What is the contractor's policy and organization for health, safety and environment issues? Does the contractor have formalized safety management procedures? What are the contractor's accident and injury statistics?

4.1.4 To what extent would work be delegated to subcontractors? How much equipment would have to be leased? What is the contractor's experience handling subcontractors in various disciplines? What types of contract and methods of reimbursement would be used?

4.1.5 Does the contractor have special procedures for dealing with identified hazards?

4.1.6 Is the contractor aware that health, safety, and environmental performance will be formally monitored by company personnel?

Main subcontractors should, as far as possible, be nominated by the prime contractor before a bid is awarded. Although company site management may not have direct control over the selection of subcontractors or of special equipment, they have a duty to their employees and to all who are on the location or live nearby to ensure safe working practice. To this end, site management should monitor the safety performance of all contractors' and subcontractors' personnel and take appropriate action to ensure that any dangerous practices or deficiencies are corrected, (see Section 5). Should there be an unavoidable change of subcontractors during the progress of the contract, the prime contractor should be required to inform location management in advance and discuss any potential problems that may arise.

Statutory requirements may exist for regulating operations such as the use of explosives or underwater work, and site management should be prepared to monitor the observance of such regulations by the contractor or subcontractor.

It is important that all staff employed by the prime contractor and subcontractors be informed of the company location safety regulations, emergency procedures, and any particular hazards that may be known to company location staff before work starts.

4.2. Work Procedures

Once the prime contractor has been selected, proposed work procedures should be discussed in detail with company site management before the contract is actually signed. The matters to be covered should include at least the following:

4.2.1 Types of equipment to be used, certification, maintenance, and periodic checking of condition.

4.2.2 Methods for cutting process piping, vessels, and tanks. Consideration should be given to the use of hydraulic shears or other cold cutting equipment to avoid residual fire risks. (This approach may also have the benefit of increasing productivity.)

4.2.3 Use of other special equipment and techniques, for example, explosives, thermal lance, water jet cutting, diving, working in inert atmospheres. From where will the skills be obtained?

4.2.4 Work Permit System for access, cleaning, removal, hot work etc. The company site permit system will normally be used and the contractor should propose a procedure for complying with this system.

4.2.5 Isolation of the plant and services will usually be carried out by company site operating and maintenance staff, and a system for identifying what needs to be isolated should be agreed on with the contractor.

4.2.6 Will testing for hazardous materials and substances be carried out by company site staff, or will third-party analysis services be required (e.g., for asbestos counts)? How will the disposal of materials be controlled?

4.2.7 Job responsibilities, terms of reference, and reporting responsibilities for the contractor's site management and supervisors: Who will meet with the company project manager and the company health, safety, and environmental specialist? Responsibility for any necessary liaison with regulatory authorities and other third parties should also be established.

4.2.8 The contractors' systems for supervising the safety of workers and operations, including subcontractors.

4.2.9 Responsibilities for reporting and investigating accidents and near misses.

4.2.10 Responsibilities and methods for testing the safety of atmospheres and for inspecting confined spaces before entry. Site management will usually wish to control these tests, and the contractor will be required to comply with the related site safe work permit system.

4.2.11 Procedures proposed by the contractor for the demolition of structures. Particular hazards should be anticipated in the felling of chimney stacks, columns, and other tall or heavy items, and in the demolition of pre- and poststressed concrete structures. Special advice may be needed in some of these areas.

4.2.12 Requirements for site orientation and other training courses for contractors' employees (e.g., operation of the safe work permit system, use of protective equipment, breaking apparatus).

4.2.13 Agreement on a way to identify piping vessels, tanks, etc, that have either been put into a safe state for further work or that remain unsafe.

4.2.14 Use of personal protective equipment (PPE). The contractor should normally provide PPE for direct and subcontracted employees. The type of equipment proposed must be acceptable to company site management.

4.2.15 Proposed methods of moving components around the site, safe access routes for cranes, heavy trucks, etc.

4.2.16 Proposed methods for steaming out, or other residual waste removal techniques, and handling and disposing of waste.

4.2.18 Methods for cutting up scrap metal and a provision of suitable areas for this work. Particular attention should be paid to the potential hazard posed by steel plate or piping that has been in contact with lead compounds.

4.2.19 The provision of temporary services, adequate lighting, telephones, power, water, sewage disposal.

4.2.20 The provision of fire prevention, detection and extinguishing systems, the location of fire alarms and fire points.

4.2.21 Ways to define the demolition site boundaries and to control the access of people and mobile construction equipment. In many cases, access will be adjacent to or even through operating plant areas, and effective control or segregation by fencing will be needed.

4.2.22 Proposals for dealing with buildings, with reference to removing glass and any other hazardous substances, to the collapse of floors, to the need for heavy lifts, etc.

4.2.23 Methods of sealing off drains, etc., that may still be capable of releasing toxic or flammable materials.

4.2.24 Proposals for using existing plant equipment such as hoists or bridge cranes. Decommissioned equipment should not be used unless current certificates of examination are available.

4.2.25 Procedures for dealing with emergencies, including site evacuation arrangements, alarm signals, and assembly areas.

4.2.26 Provisions for first aid and medical facilities.

After these discussions, site management should require the contractor to submit a draft detailed plan of the sequence of demolition, summarizing the proposed procedures.

4.3. Modifications

After demolition begins, technical or commercial pressures may oblige the contractor to modify the original plan of work. The terms of the contract should make clear that it is the prime contractors' responsibility to inform site management in advance of any proposed change to the plan. Site management should then review and approve the change, if it is acceptable. Consultation with key operating and maintenance staff is crucial to the review process.

5.0 MONITORING HEALTH AND SAFETY DURING THE PROJECT

5.1. Commencement of Contract

Immediately before the start of the demolition work, it is advantageous to arrange a meeting between company location management, the contractors' site representatives, and, when appropriate, representatives of the local regulatory agency(ies). The meeting should address the proposed plan of work and the essential features of process safety management during the contract. Such a meeting will help emphasize to the contractor the importance that site management places on the observation of good process safety practice throughout the work.

5.2. Coordination Meetings

Once demolition has begun, coordination meetings should be held between company location management and the contractor at least once a week, during which the progress of the work can be monitored. Any health, safety, or environmental problems that have arisen should be discussed and resolved. The proposed work plan for the forthcoming week should be reviewed and relevant process safety aspects highlighted. Brief minutes of these meetings should be issued to all concerned as quickly as practicable to ensure that agreed-on actions are followed through.

5.3. Contractual Obligations

As well as holding regular coordination meetings, company site management should ensure that corporate interests are safeguarded by formally monitoring the contractors' health and safety performance. The following matters should be addressed, either by written agreement with the contractor or through the contract document itself.

5.3.1 The right of access to the site by company representatives.

5.3.2 Frequency of safety reviews and methods of reporting.

5.3.3 Basis for company location management halting work.

5.3.4 Procedures for resolving disputes about health, safety, and environmental matters.

5.3.5 Clarification of the responsibilities of all parties involved for dealing with emergencies.

When appropriate, the contract document should also incorporate provisions to enable the company to terminate the engagement of the contractor, or to terminate contractor or subcontractor employees in the event of unsatisfactory safety performance.

5.4. Documentation

Company location management should require documentary evidence of the following:

5.4.1 Certificates of periodic examination of cranes, lifting tackle, air receivers, and other critical equipment must be current and valid, particularly where regulatory requirements exists.

5.4.2   Mobile construction equipment drivers should be trained and experienced.

5.4.3   The reporting of accidents and incidents.

5.4.4   The adequacy of personal monitoring of those exposed to toxic or radiation hazards (e.g., for lead, asbestos, X-ray), including, when appropriate, evidence of satisfactory initial medical examination.

5.5. Other Aspects

Following are further examples of contractor's activities that should be monitored:

5.5.1   The correct use of personal protective equipment.

5.5.2   The safe treatment and disposal of hazardous substances.

5.5.3   The observance of agreed-on working hours. Changes of shift should be coordinated so that adequate safety, medical, and fire cover is always available.

5.5.4   The correct handling of debris, fumes, spillages, etc. and ensuring that personnel in adjacent areas are not at risk or unprotected.

5.5.5   The correct use of scaffolding, ladders, chutes, cranes, hoists, mobile plant, cutting equipment.

5.6. Conclusion of Project

At the end of the demolition operations, a final safety review should be carried out to determine whether the location is in an acceptably safe state to allow it to be handed over to a new owner or to be left without further attention. Careful supervision of the contractor during the final phases of the project will be required to ensure that safe conditions are finally established before the contractor leaves the site.

C

EXAMPLE OF SITE-SPECIFIC DEMOLITION CHECKLIST/QUESTIONNAIRE

1.0. REVIEW OF THE ADEQUACY OF THE PLANT PREPARATIONS

1. Who has surveyed the site, examined the condition of the buildings and structures? Has a report been written indicating problem areas? Do any equipment/structures require further special advice about the method of removal?
2. Are floors, stairways, etc. safe to use?
3. What liaison has been established among all the parties concerned, i.e., the Safety and Health Department of the contractors, the demolition inspector, and XYZ plant personnel? Will there be inspections during various stages of demolition to ensure that safe practices are in fact being carried out?
4. Is the area for the demolition completely isolated? Are piping and services (electricity, etc.) disconnected and/or blinded? Who has made the physical check? Who has checked any rerouted lines, etc.? Are there Plant Modification Sheets? Are there any temporary live services in the area? Are these clearly identified?
5. What checks have been carried out of underground hazards in the area? Does process area "A" drainage pass through process area "B"? What is underneath the soft ground where cranes, etc. are likely to stand?
6. Has a comprehensive list of chemicals, including intermediate compounds, processed or stored in the areas in the present or past been established? Is this list accompanied by all relevant flammable and toxicity data, etc.?

7. Is a safe work permit system in force for opening up each piece of equipment and piping for cleaning? Will there be a separate permit for actually removing each piece of equipment or piping?

8. Can all vessels and piping that have been cleaned, etc. be easily identified? Is there a clear identification system (color code) for equipment and piping to indicate its safety status? For example, may it be removed to the steaming out point?

9. Do any of the vessels have separate linings that could create process hazards?

10. Has the process supervisor responsible for issuing "B" type permits enough knowledge of the buildings and chemicals handled in them? Who will accept the permits? The contractor or company inspector?

11. Where will the demarcation come between work carried out by XYZ Plant maintenance personnel and work carried out by the contractor?

12. Has all the asbestos lagging been removed?

13. Will safety showers and eye wash bottles be provided in the area?

14. Has a study been made of previous dangerous occurrences and minor accidents that have occurred during the demolition of similar process equipment? Will work procedures prevent similar incidents from happening again?

15. Will safety boots, helmets, gloves and goggles be standard issue to all contractor personnel? Will the wearing of this safety gear be mandatory?

16. Will safety harnesses, etc. be available at all times on the site?

17. Will the steaming out point be clearly segregated from the cutting and breaking area to prevent the possibility of confusion? How will equipment be moved from one area to another?

18. Where will the acetylene, propane, and oxygen bottles be stored? Who checks the flexible hoses on the equipment?

19. What steps have been taken to remove all the materials at present stored in the demolition area? Some of the materials are hazardous, e.g., sulfuric acid in carboys, hypochlorite solutions in carboys, and D.F. esters in the tanks to the east of Process Area "A".

20. Has the paint been tested for lead content? Will the demolition contractors' employees be given medical checks if necessary? Who decides?

21. Has the fire prevention officer been consulted about fire points, alarm points, hydrants, etc.? Is he or she satisfied that sufficient access is available for emergencies? Will fire station personnel make an inspection of the site at the end of each working day?

22. Has the electrical classification for the surrounding areas been considered for the selection of the site for hot work, etc.?

23. Is the lighting adequate for the hours of work?

24. Is the demolition area adequately signposted?
    a. Danger Demolition Area
    b. Danger Asbestos Stripping Ongoing

      c. Demolition Traffic Only

      d. Demolition Control Office

25. Have the contractor employees been allocated a building, etc. for keeping their property safe?

26. Where will equipment to be recovered and claimed by other company personnel be stored until required?

27. Are any special preparations needed for dismantling windows and cladding? Could the cladding be made of cement or asbestos?

28. Will equipment and structural drawings be available to the contractor and demolition inspector to identify loading within each building and the safe approach for dismantling?

29. Are any vessels or piping yet to be cleared of toxic or flammable chemicals? What special arrangements have been made for these items?

30. If certain items are being gas freed while the contractor is dismantling, what safeguards are being undertaken to prevent the exposure of workers to fumes?

31. How will the shift plant managers be kept informed of daily progress and problems?

32. Is a telephone available to the contractor for calls within the plant or outside?

2.0. REVIEW OF THE WORK METHODS AND SUPERVISION

1. Has the demolition inspector a through knowledge of demolition work and also of the principles of building construction?

2. Is a chart available showing the process safety management system (e.g., safe work permits and the coordination needed among those involved in the demolition)? Will a responsible plant maintenance supervisor be available at all times on the site?

3. Does a list exist of personnel involved in the demolition? Will there be any need for access to the demolition? Will there be any need for access to the demolition area by persons other than those involved? Will such entry be controlled? What happens if one of the demolition specialists is off work?

4. Do all personnel involved in the demolition fully understand the process safety management system in force at the XYZ Plant?

    Safe Work Permits   Process Safety Information

    Hot Work Permits   Emergency Planning and Response

    Management of Change   Contractor Safety

    Are copies available on site? Will a special alarm system be provided for use in the demolition area?

5. Has a consultant structural engineer been appointed? Is there a procedure for obtaining this type of expert advice?

6. Are XYZ Plant supervisors and the contractor conversant with the following regulatory standards?
   a. OSHA Demolition Standard 29 CFR 1926.850
   b. OSHA Crawler and Truck Crane Standard 29 CFR 1910.180
   c. OSHA Material Handling Standard 29 CFR 1910.176
   d. OSHA Construction Work Standard 29 CFR 1910.12
   e. OSHA Means of Egress Standard 29 CFR 1910.37
   f. OSHA Scaffolding Standard 29 CFR 1910.29
   g. OSHA Medical Services and First Aid Standard 29 CFR 1910.151
   h. OSHA Cutting and Welding Standard 29 CFR 1910.252
   i. OSHA Asbestos Standard 29 CFR 1910.1101
7. Who has checked to ensure that all these regulations are met? For example, restroom facilities, medical (first aid) facilities, etc.
8. Should all the relevant state and federal regulations and the XYZ Plant standards be written into the contract?
9. Is the contractor adequately insured to cover all possible contingencies?
10. Has the contractor employed competent personnel, for example, a sling hand for crane work (thorough knowledge of signals, etc.), a certified crane operator with previous demolition experience?
11. Is there a master plan showing the sequence of demolition? Will the results of each step be forecast accurately? Who is responsible for each step? Who is in charge overall?
12. Who will ensure that all construction equipment on site is in good condition and has been regularly inspected to meet all XYZ Plant and regulatory requirements?
13. What is the standard of housekeeping expected to be on a job of this kind? Have safe access, a means of escape, tripping hazards, holes in flooring, etc. been taken into account? Will any doorways be allocated safe for entry/exit? Will they be protected?
14. Will all the windows and side cladding be removed first?
15. Will any flame cutting take place inside the buildings? When within the scope of work will this be done? Are there potential sources of flammable materials inside the buildings?
16. Will there be a set procedure for lowering large items of equipment? Will it be necessary to meet with other adjacent area and XYZ Plant shift managers?
17. Will "long term" permits be issued for any purpose? If so, where and for what purpose?
18. Who will renew permits each morning to allow work to progress safely and without delay?
19. Where will the permits be kept? Does everyone understand the permit system?
20. Has safety clearance been given for using hoists within the building?

21. When "C" type permits are issued, how will the contractors differentiate between equipment for:
    a. Scrap
    b. Removal for sale
    c. Retainment for use by the company at other locations
22. Is it possible to overload a truck with scrap? Will it pass a weigh station?
23. What demolition methods are going to be used on tanks and vessels?
24. What "incident" or "accident" documentation will be held by:
    a. XYZ Plant
    b. Contractor
25. Have the company's head office health, safety and environmental staff been informed of this work? Will they be making regular inspections?
26. Will gas tests for flammability be conducted on vessels before removal from buildings and again before doing hot work?
27. Because it will aid the continuity of work, will preliminary atmosphere tests (flammability, toxicity) be made by the XYZ Plant operations shift supervisor before any unit or equipment is isolated? Such tests will help determine the care needed when isolating units or equipment and when preparing the plant (steaming out). Obviously, a plant laboratory test must be done before any hot work or entry situation.
28. Will grinding wheels be used? If so, has a competent person been assigned to check and change the wheels?

3.0. STUDY OF THE EFFECTS OF THE WORK ON THE SAFETY OF OTHER XYZ PLANT ACTIVITIES

1. What effect will the demolition have on surrounding units, filling stations, railways, tank farms, etc.?
   e.g,
    a. Process Area "B"
    b. Process Area "C"
    c. Tank Farm
    d. Rail Loading Dock
    e. Pipeline Pump Station
2. What effect will the area electrical classification for the surrounding areas have on the demolition program?
3. How will personnel working around the area be protected from falling debris?
4. Will scrap be removed off site as soon as a "C" type permit has been issued? Will the "C" type be used as a "Security Pass Out" at the main gate, or will other documents be required, with an XYZ Plant signatory proving inspection and certification of the load?
5. How will you prevent access to fire fighting equipment being blocked with scrap?

6. Roadways in the area are very narrow. What measures are anticipated to keep them clear? Will traffic not involved with demolition be re-scheduled to outside normal working hours?

7. Will the demolition of Process Area "A" affect loading at the truck loading dock?

8. Will all "A" type permits in the area be countersigned by operations shift supervisors responsible for the units, etc. in the surrounding areas?

9. The railway line and pipe racks surround the north and east areas of demolition. How will they be protected against falling objects, etc.?

10. Has the Materials Handling Section been informed of the demolition since it will affect their operations in this area?

11. How will the demolition affect personnel working in the waste water disposal area?

12. Will the demolition area be fenced off in one large area or will parts of the working area be fenced off separately (e.g., steaming point, scrap collection, and rubble collection areas)?

13. Will the movement of any materials (e.g., recovered chemicals from vessels placed in drums) be likely to expose persons inside the plant? If the drums are disposed of outside of XYZ Plant, will the general public be exposed to any hazards?

14. Will vehicle loads be inspected before leaving the XYZ Plant?

15. Will all asbestos-containing material leaving the XYZ Plant be suitably packaged and labelled? Will it be disposed of at an approved site?

16. Will any materials such as rubble be disposed of within the XYZ Plant?

17. Will the routes for the contractors' vehicles be clearly defined within the plant?

INDEX

Troubleshooting maintenance, described, 213–214
Turnaround, startup after, 132–134
Turnkey project, pre-startup, organization and roles, 76

U
U.S. Department of Labor, 2–3
U.S. Environmental Protection Agency (EPA), 3
Unit restart, normal shutdown steps, 259
Upgrade training, maintenance training, 225–226
Utilities, commissioning of, 105

W
Wear, aging equipment maintenance, 218–219
"What-if" games, operator training, 180, 183
Work order tracking, maintenance, management information systems, 232–233
Work permits, maintenance, 226–232
Written procedures
  normal shutdown steps, 255
  routine operations, 143–146. *See also* Operating procedures